NUCLEAR MAGNETIC RESONANCE IN A FLOWING LIQUID

YADERNYI MAGNITNYI REZONANS V PROTOCHNOI ZHIDKOSTI

ЯДЕРНЫЙ МАГНИТНЫЙ РЕЗОНАНС В ПРОТОЧНОЙ ЖИДКОСТИ

Nuclear Magnetic Resonance in a Flowing Liquid

by
Aleksandr Ivanovich Zhernovoi
and
Georgii Dmitrievich Latyshev

Authorized translation from the Russian by
C. Nigel Turton, B.Sc., Ph.D., and Tatiana I. Turton, B. A.

Springer Science+Business Media, LLC
1965

The Russian text was published by Atomizdat in Moscow in 1965.

Александр Иванович Жерновой,
Георгий Дмитриевич Латышев
ЯДЕРНЫЙ МАГНИТНЫЙ РЕЗОНАНС В ПРОТОЧНОЙ
ЖИДКОСТИ

Library of Congress Catalog Card Number 65-13580

ISBN 978-1-4899-4925-7   ISBN 978-1-4899-4923-3 (eBook)
DOI 10.1007/978-1-4899-4923-3

PREFACE

At the present time there are several books in which the theory and application of nuclear magnetic resonance in stationary media are examined systematically.

The phenomenon of nuclear resonance in a flowing liquid has been considered only in scattered articles and then very incompletely.

In writing this monograph the authors undertook to give an account of the peculiarities of nuclear magnetic resonance in a flowing liquid on the basis of available literature material and their own unpublished work. In the introduction there is a review of the main literature data, in the first part the results of theoretical and experimental investigations of a flow detector are examined, and in the second part the possibilities of its practical application are considered.

The use of nuclear magnetic resonance in a flowing liquid makes it possible to carry out contactless measurement of liquid flow, analysis of the composition and relaxation time of a substance flowing continuously in a tube, the investigation of turbulent mixing processes, the measurement and stabilization of weak magnetic fields, etc.

The book is intended for scientific workers, graduate students, engineers, and students specializing in the practical application of nuclear resonance.

CONTENTS

INTRODUCTION

Nuclear magnetic resonance in a flowing liquid was first observed by the Indian scientist Suryan [1]. In his experiment the detector was a radio-frequency coil wound on a glass tube through which passed an aqueous solution of $FeCl_3$. The tube was placed in a strong uniform magnetic field and the signal was detected with a bridge circuit. Suryan observed that if the solution concentration gave a proton relaxation time within the range of 0.1-0.05 sec, the amplitude of the proton resonance signal A in the moving liquid was considerably greater than the amplitude A_0 in the stationary liquid.

The experimental relation of the signal amplitude to the liquid flow rate W obtained by Suryan for 0.001 N $FeCl_3$ solution in water ($T_1 = 0.05$ sec) is given in Fig. 1.I. The increase in the signal amplitude in the moving liquid is explained by the influx into the detector of polarized nuclei, appearing when the liquid flows into the magnetic field before entering the detector. Let us examine this effect from the point of view of the theory used normally in the literature.

The amplitude of the nuclear resonance signal is proportional to the magnetization of the nuclei, i.e., the total magnetic moment of the nuclei in unit volume of the liquid in the detector. A change in the magnetization of the nuclei M occurs as a result of two processes: 1) the interaction of the nuclei with the resonance oscillating field, whereby the magnetization falls with a rate

$$\frac{dM}{dt} = -\gamma^2 H_1^2 T_2 M, \tag{1.I}$$

where γ is the gyromagnetic ratio of the nuclei, H_1 is half the strength of the resonance oscillating field in the detector, and T_2 is the reciprocal half-width of the nuclear resonance line in frequency units; 2) the polarization of the nuclei in the detector by the external magnetic field H_0, whereby the magnetization increases at a rate

$$\frac{dM}{dt} = \frac{X_0 H_0 - M}{T_1}, \tag{2.I}$$

where X_0 is the static nuclear magnetic susceptibility, H_0 the strength of the external magnetic field in the detector, and T_1 the longitudinal nuclear relaxation time.

As a result of the two processes, the magnetization changes with time at a rate

$$\frac{dM}{dt} = -\gamma^2 H_1^2 T_2 M + \frac{X_0 H_0 - M}{T_1}. \tag{3.I}$$

When the substance in the detector is stationary, the magnetization of the nuclei M_0 has the same value throughout the whole of its volume. As the nuclear magnetic resonance process is in a steady state, then $dM_0/dt = 0$, and, consequently, from expression (3.I) we have

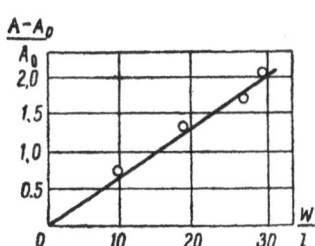

Fig. 1.I. Experimental relation of the amplitude of the nuclear absorption signal in the flow detector to the flow rate of a 0.001 N solution of $FeCl_3$ in water [1] (l is the length of the detector coil).

1

$$M_0 = \frac{X_0 H_0}{1 + \gamma^2 H_1^2 T_1 T_2} = X_0 H_0 Z, \qquad (4.\text{I})$$

where Z is the saturation factor.

Thus, with a stationary liquid the signal amplitude A_0 is proportional to $X_0 H_0 Z$. If the liquid is moving through the detector, at the entrance to the latter it has magnetization of the nuclei M_{en}, which is independent of the saturation factor. In the movement of the liquid along the detector the magnetization M falls at a rate determined by expression (3.I). By replacing dM/dt by (dM/dx)W (where x is the coordinate along the stream in the detector and W is the liquid flow rate) and integrating with respect to x we obtain the distribution of M along the detector.

$$M(x) = (M_{en} - X_0 H_0 Z)\, e^{-\frac{x}{W T_1 Z}} + X_0 H_0 Z. \qquad (5.\text{I})$$

The signal amplitude A is proportional to the value of M averaged over the length of the detector l:

$$A \sim \overline{M} = \frac{1}{l} \int_0^l M(x)\, dx = (M_{en} - X_0 H_0 Z)\frac{W T_1 Z}{l}\left(1 - e^{-\frac{l}{W T_1 Z}}\right) + X_0 H_0 Z. \qquad (6.\text{I})$$

The relative increase in the signal amplitude with motion of the liquid

$$\frac{A - A_0}{A_0} = \frac{M_{en} - X_0 H_0 Z}{X_0 H_0}\left(1 - e^{-\frac{l}{W T_1 Z}}\right)\frac{W T_1}{l}. \qquad (7.\text{I})$$

In Suryan's experiment the liquid flowed in the magnetic field H_0 for a sufficient time for equilibrium polarization of the protons before entering the detector, i.e.,

$$M_{en} = X_0 H_0,$$

while the value $Z \ll 1$. Under these conditions, the expression (7.I) is simplified to

$$\frac{A - A_0}{A_0} = \frac{W T_1}{l}. \qquad (8.\text{I})$$

Thus, the slope of the line (see Fig. 1.I) relative to the abscissa axis is proportional to the relaxation time of the liquid. On this basis Suryan proposed the use of a flow detector for measuring the relaxation time of liquids in the range of 0.1-0.05 sec.

Suryan also observed that in some cases a considerable increase in the signal amplitude is observed only in the first moment after the beginning of movement of the liquid and then it falls to some steady value which may be insignificantly greater than the amplitude in the stationary liquid. This effect is produced by incomplete magnetization of the nuclei in the liquid while it is flowing in the field H_0 before entering the detector. When the liquid is stationary, the magnetization of the nuclei in the field H_0 outside the detector coil is able to reach the equilibrium value $X_0 H_0$ and at the beginning of motion, nuclei with a high magnetization pass into the detector and this produces a considerable increase in the signal amplitude. When the polarized liquid has passed through the detector, into the latter there begin to flow nuclei with a considerably lower magnetization M_{en}, which arises during the brief flow of the liquid in the field H_0 before entering the detector. Thereupon the signal amplitude falls. If $M_{en} < X_0 H_0 Z$, the signal amplitude in the flowing liquid is less than in the stationary liquid.

Denis, Béné, and Exterman [2] also observed an increase in the signal amplitude with movement of the liquid and explained it by acceleration of the relaxation of the mean magnetization of the nuclei through the volume of the detector, which occurs as a result of the influx of liquid with equilibrium magnetization into the detector.

They concluded that the signal amplitude in a flow detector corresponds to the effective relaxation time T_1'', which is related to the true relaxation time of the liquid T_1 by the expression

$$\frac{1}{T_1''} = \frac{1}{T_1} + \frac{1}{T_1'} , \qquad (9.\text{I})$$

where T_1' is the time after which $1/e$ of depolarized nuclei is replaced by polarized nuclei as a result of the flow of liquid into the resonance zone. This relation may be understood by examining the restoration of the nuclear magnetization vector in the detector.

If in the liquid entering the detector the nuclei have a magnetization M_{en} and in the liquid flowing out, M_{ex}, then as a result of the flow of the liquid at a rate q the mean magnetization of the nuclei M through the detector volume v changes at a rate

$$\frac{dM}{dt} = \frac{M_{en} - M_{ex}}{v} q. \qquad (10.\text{I})$$

The change in M as a result of spin-lattice relaxation is described by the expression

$$\frac{dM}{dt} = \frac{M_0 - M}{T_1} , \qquad \cdot$$

where M_0 is the equilibrium magnetization in the detector. With the condition that $M_{en} = M_0$ and $M_{ex} = M$, the total rate of change of M is determined by the expression

$$\frac{dM}{dt} = (M_0 - M)\left(\frac{1}{T_1} + \frac{q}{v}\right) = \frac{M_0 - M}{T_1''} . \qquad (11.\text{I})$$

Thus, if the liquid is completely polarized before entering the detector and then rapid mixing occurs in it, expression (9.I) holds and $T_1' = v/q$. In other cases T_1' is a complex function of the liquid flow rate and the saturation factor in the detector.

Bloom and Shoolery [3] studied experimentally the possibility of using a flow detector in nuclear resonance spectrometers with a high resolution for artificially reducing the longitudinal relaxation time by changing the working substance. They observed that in the investigation of chemical signals in simple organic substances there is an appreciable improvement in the signal-to-noise ratio, but the advantage of a flow detector in the investigation of spin-spin interaction is doubtful.

Sherman [4] carried out an experiment with a nuclear induction flow detector. The exciting and receiving coils lay at a distance of 6 mm from each other and water passed through them successively. A signal was observed with the frequency of the exciting field and not the precession frequency and its amplitude increased with an increase in the liquid velocity. A similar investigation of this effect was carried out by the author in later work [5], where it was shown that by this method it is possible to measure the mean strength of the field in the section of tube between the exciting and receiving coils.

Gaussen [6] proposed a method of measuring long relaxation times T_1 by means of a flow detector. The method was as follows: the liquid first passed through the nuclear resonance detector at a high rate without being magnetized and therefore there was no signal, then at a certain moment of time the liquid was suddenly stopped and from this moment there began polarization of the nuclei, accompanied by a corresponding increase in the signal. By recording the change in signal with time on a movie film it is possible to determine the relaxation time of the liquid.

Mitchell and Phillips [7] used a flow detector in an apparatus for determining the concentration of H_2O in heavy water. The long relaxation time of water greatly reduces the signal amplitude and this hampers the measurements, especially at low concentrations. Therefore, for reducing the effective relaxation time, at low concentrations the signal amplitude was measured with a flowing liquid and this considerably increased the accuracy.

Hrynkiewicz and Waluga [8] built an apparatus for measuring relaxation time by Suryan's method. The detector was a radio-frequency coil 15 mm long, wound on a tube 7 mm in diameter. It was placed in a magnetic field with a strength $H_0 = 6580$ oe, produced by an electromagnet with a pole diameter of 140 mm and a gap of 30 mm. The nonuniformity of the field in the detector volume was 0.02 oe. The length of the tube placed in the magnet gap through which the liquid flowed to the detector was 360 cm. The liquid flow rate was varied by connecting into the system one of 20 calibrated capillaries. To measure the relaxation time T_1 it was necessary to obtain the relation of the nuclear magnetic resonance signal amplitude to the liquid flow rate W. One of these relations for the protons of distilled water is given in Fig. 2.I. The maximum water flow rate was 40 cm/sec and consequently the time it spent in the polarizing field before entering the detector was 9 sec, which was sufficient for complete polarization of the nuclei, i.e., the magnetization of the protons in the water entering the detector $M_{en} = X_0 H_0$. Then from expression (7.I) we obtain the relation of the signal amplitude to the liquid flow rate W

$$\frac{A - A_0}{A_0} = (1 - Z)(1 - e^{-\frac{l}{WT_1 Z}}) \frac{WT_1}{l}, \qquad (12.I)$$

where A_0 is the signal amplitude at $W = 0$. At low flow rates when $l/WT_1 Z \gg 1$, this expression is simplified:

$$\frac{A - A_0}{A_0} = (1 - Z) \frac{WT_1}{l}. \qquad (13.I)$$

Suryan proposed the determination of T_1 by using this relation at $Z \ll 1$. As considerable saturation makes it difficult to obtain a satisfactory signal amplitude at long relaxation times, this method is limited to an upper limit for the measurement of T_1 of 0.1 sec.

Hrynkiewicz and Waluga refined Suryan's method. They proposed the use of relation (12.I) for finding T_1 at low saturation, when it is necessary to know the value of Z. It may be determined from the signal amplitude A_∞ at a high liquid flow rate when $l/WT_1 Z \ll 1$. In this case the expression for the signal amplitude (12.I) has the form

$$\frac{A_\infty - A_0}{A_0} = \frac{1 - Z}{Z}. \qquad (14.I)$$

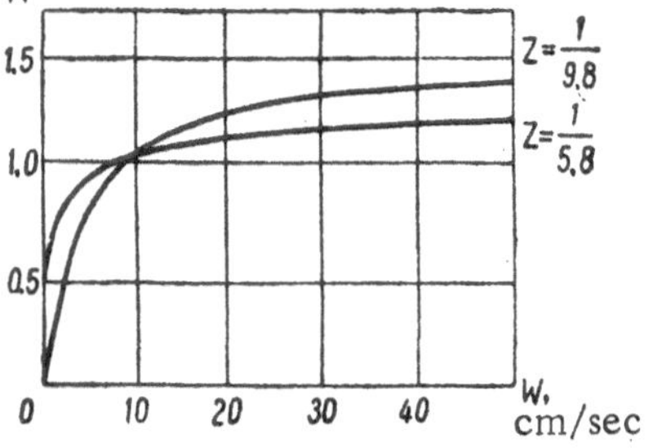

Fig. 2.I. Experimental relation of the nuclear absorption signal amplitude in a flow detector to the water flow rate [8].

By determining from the graph in Fig. 2.I the signal amplitude A_0 at zero liquid flow rate and the value A_∞ to which the signal amplitude tends with an increase in the liquid rate, from expression (14.I) it is possible to find the saturation factor Z and from the slope of the linear dependence of the signal amplitude on W at a low liquid flow rate, the value of $T_1(1 - Z)$ in accordance with expression (13.I).

Thus, this method may also be used when $Z \approx 1$ and this makes it possible to measure T_1 of the order of several seconds.

Yet another method of measuring the relaxation time T_1 by means of a flow detector was proposed by Antonowicz [9, 10]. In his apparatus a flow detector with two coils was used, i.e., the signal was detected by Bloch's induction method. The receiving coil was wound directly on the tube through which the liquid investigated flowed, while the transmitting coil lay perpendicular to it. The section of the tube immediately before the receiving coil was carefully shielded from the oscillating magnetic field produced by the transmitting coil with a brass screen. The relaxation time was measured in the following way. At the initial moment the liquid was stationary and a nuclear resonance signal weakened by the saturation effect was recorded by the detector. In the shielded section of the tube before the detector the liquid was polarized, reaching equilibrium magnetization of the nuclei. When the liquid began to move, this polarized liquid entered the detector, producing a brief marked increase in the signal amplitude to a certain value A_0, as was observed by Suryan. When all the equilibrium polarized liquid had passed through the detector, liquid with a lower magnetization began to enter the detector and the signal amplitude fell.

An approximate examination of the theory of this effect led Antonowicz to the conclusion that at a given liquid flow rate the maximum value of A_0 is observed at some optimal amplitude of the oscillating field H_{1opt} in the transmitting coil. This optimal amplitude is determined by the condition

$$\gamma^2 H_{1opt}^2 T_{1eff}\, T_{2eff} = 1,$$

where

$$T_{1eff} = \frac{T_1 T_0}{T_1 + T_0}, \quad T_{2eff} = \frac{T_2 T_0}{T_2 + T_0}, \quad T_0 = \frac{l}{W},$$

l is the length of the receiving coil. Normally $T_2 \ll T_0$ and then $T_{2eff} = T_2$. Having established and measured H_{1opt}' and H_{1opt}'' at two liquid flow rates W_1 and W_2, respectively, it is possible to find T_1 from the expression

$$\frac{T_1 + \dfrac{l}{W_1}}{T_1 + \dfrac{l}{W_2}} = \frac{W_1}{W_2} \left(\frac{H_1'' \, _{opt}}{H_1' \, _{opt}} \right)^2. \tag{15.I}$$

It is readily seen that the theory of the method is based on the validity of expression (11.I), i.e., it is correct if the magnetization of the nuclei does not change along the detector. This was not provided for in Antonowicz's apparatus.

In all the work examined above, the polarization of the liquid and the detection of the nuclear resonance signal were carried out in the same strong magnetic field.

In 1957, for the first time the polarization of the liquid and the detection of the nuclear resonance signal in a flow detector were carried out in different magnetic fields. This was done independently by A.I. Zhernovoi for developing a method of measuring and stabilizing a weak magnetic field and by F. I. Skripov to measure the earth's field. In the apparatus designed for Zhernovoi's method, water flowed for a certain time through a space in a magnetic field of the order of 5000 oe produced by a polarizing electromagnet; it then passed rapidly through a connecting tube into the nuclear resonance detector in a weak magnetic field, where the absorption signal was recorded by means of an autodyne detector [11, 12]. In F. I. Skripov's work [13] the liquid was polarized with the field of a solenoid with a strength of about 100 oe, while the nuclear resonance detector was in the earth's magnetic field, where the free precession signal was recorded.

If the polarization of the nuclei and the detection of the nuclear resonance signal are carried out in the same field H_0, the signal amplitude is proportional to the square of the field strength as it is proportional to the excess of nuclei in the lower energy state, i.e., the magnetization of the nuclei $M = X_0 H_0$ and also the energy which is absorbed by the nuclei on transition from one energy level to the other $\Delta E = h\gamma H_0$. With preliminary polarization of the flowing liquid the magnetization inside the detector is independent of the strength of the field in it H_0, but is proportional to the strength of the polarizing field H_{pol}, and therefore the signal amplitude is proportional to the product $H_{pol} H_0$.

The weakening of the dependence of the signal amplitude in a flow detector with preliminary polarization on the strength of the external field makes it possible to obtain a nuclear magnetic resonance effect in weak magnetic fields. For example, by using a flow detector it was possible to measure and stabilize a magnetic field with a strength of a few oersteds with a detector volume of 0.03 cc [11, 12, 14] and to obtain a continuous signal of free precession in the earth's field, which is proposed for use for geomagnetic measurements [13]. A flow detector has also been used for building spectrometers for studying nuclear magnetic resonance in a weak field [15-17].

One of the most interesting characteristics of a flow detector is the possibility of obtaining in it a continuous flow of liquid with negative polarization of the nuclei. This effect was first obtained by A. I. Zhernovoi in developing a method of measuring a weak nonuniform magnetic field, by Benoit in producing an auto-oscillating system, and by Wilking in studying multiquantum transitions. Liquid flowing from the polarizing field into the nuclear resonance detector was subjected to a resonance oscillating field produced by a radio-frequency coil wound on the tube. When the nuclei passed through this coil their magnetic moments were flipped relative to the direction of the external field; as a result, the magnetization of the nuclei in the liquid emerging from the coil was negative. On entering the nuclear resonance detector, the negatively polarized liquid gave a nuclear emission signal instead of an absorption signal. This effect, which is called the "nutation effect," has been described in several papers [18-25]. The degree of flipping of the nuclei produced by the oscillating field is independent of the strength of the external field and depends little on its nonuniformity. On this basis it is possible to measure magnetic fields over a wide range of strengths by placing a reversing coil in them and fixing the resonance frequency of the oscillating field.

As has already been mentioned, when a negatively polarized liquid enters the nuclear resonance detector, an emission signal is observed instead of an absorption signal in it, i.e., the nuclei radiate energy into the radio-frequency circuit. Under certain conditions, in particular, if the Q-factor of the circuit and the magnetization of the nuclei M are sufficiently great, the nonuniformity of the external magnetic field in the circuit is low, and the natural frequency of the circuit is close to the precession frequency of the nuclei in the external field, then the system nuclei-resonance circuit is unstable and the inflow of nuclei into the circuit excites an auto-oscillation in it with a frequency intermediate between the natural frequency of the circuit and the precession frequency of the nuclei.

Such a system, which is called a "nuclear resonance maser," was first produced by the French scientist Benoit [22-24].[1] Many papers have been devoted to the investigation and application of the "maser" [25-33]. For obtaining auto-oscillations, the Q-factor of the circuit was artificially increased by electronic methods. The oscillation frequency of the system is closer to the precession frequency of the nuclei, the smaller the Q-factor of the circuit and the more uniform the external magnetic field [25, 26]. By using a sufficiently homogeneous weak field and a detector of special design in which the liquid was rotated it was possible to obtain auto-oscillation without an artificial increase in the Q-factor of the circuit [27]. In such a "maser" the oscillation frequency is close to the precession frequency of the nuclei and this may be used, for example, for measurement and stabilization of a field [29].

A "maser" with a flow detector has been used as a nuclear resonance spectrometer. With circulation of ammonium hydroxide solution, which has a hyperfine structure of the proton resonance line, self-excitation of the "maser" was observed at each component of the spectrum [28].

The displacement and broadening of the nuclear resonance line in a flow detector, which is called the instrument effect, was described for the first time in 1960 [34]. On the basis of this effect a simple method was proposed for determining the sign of the gyromagnetic ratio of nuclei [35] and a method of measuring liquid flow rate [36]. In 1959 A. I. Zhernovoi and somewhat later and independently, Singer (U.S.A.) proposed methods of measuring the flow rate of a liquid in a tube from the time taken to move a fixed distance by nuclei labeled by nuclear resonance and from the signal amplitude [38-42]. Nuclear magnetic resonance has been used for hydrodynamic investigations by some authors [43-45]. In 1961 a method was proposed for measuring the flow rate of a liquid from the strength of the resonance oscillating magnetic field produced by nutation of the magnetization of the nuclei through a given angle [194].

---

[1]As a nuclear resonance maser operates in the radiofrequency range and not in the microwave region, it would be more correct to call it a "raser."

On the basis of a flow detector a method was developed for measuring the relaxation time in a flowing liquid [46-48].

The flow detectors used had three units:

1) a polarizer — a device in which the moving liquid acquired high magnetization of the nuclei;

2) a nuclear resonance detector which was designed for observing the signal produced by the magnetization of the nuclei of the liquid flowing from the polarizer;

3) a nutation detector — a radiofrequency coil through which the liquid passed in flowing from the polarizer to the nuclear resonance detector. The resonance oscillating field of this coil produced a deviation (nutation) of the magnetization of the nuclei of the liquid from the direction of the external field and this was recorded through the change in the signal in the nuclear resonance detector. In the next three chapters we examine the processes occurring in each of these three units of the flow detector, namely, the polarizer (Ch. 1), the nutation detector (Ch. 2), and the nuclear resonance detector (Ch. 3). In the second part of the book we describe possible applications of a flow detector and give the calculations of the parameters of the corresponding instruments. It should be noted that these calculations are mainly for the sake of the method as many of them have not been checked adequately in practice. The authors considered it necessary to present these calculations in order to illustrate the effect of various factors on the characteristics of instruments containing a nuclear resonance flow detector.

As theoretical and experimental investigation shows, in all cases of the practical application of a flow detector the spread of the velocities of the molecules in the stream produce additional errors in the measurements. These errors are much lower if the velocities are artificially leveled off across the section of the tube. Therefore, in the subsequent examination of the theory of a flow detector it is assumed that the spread of molecular velocities is small in order to simplify the discussion. The effect of this small spread is allowed for by introducing the effective "nutation" relaxation times T_{1n} and T_{2n}.

Part I

Characteristics of Nuclear Magnetic
Resonance in a Flowing Liquid

CHAPTER 1

PRELIMINARY POLARIZATION OF THE FLOWING LIQUID

1.1 Methods of Polarizing Nuclei

The magnetic moments of nuclei of a substance placed in a magnetic field may take up several definite positions relative to the direction of the field. If the nuclei have a spin of 1/2, there are two such positions: the nuclear spin may be oriented in the direction of the field or against it. In this case, the projection of the magnetic moment of the nucleus on the direction of the field equals $+\mu$ or $-\mu$. In a magnetic field H, the state in which the projection of the magnetic moment equals $-\mu$ has an excess energy of $2\mu H$ relative to the state with a projection of the nuclear magnetic moment of $+\mu$.

With complete thermodynamic equilibrium of the system of nuclei in a magnetic field, in unit volume of the substance the number of nuclei with magnetic moments oriented along the field $N(+\mu)$ exceeds the number of nuclei with magnetic moments oriented against the field $N(-\mu)$, and

$$\frac{N(+\mu)}{N(-\mu)} = e^{\frac{2\mu H}{kT}},$$ (1.1)

where k is the Boltzmann constant and T the absolute temperature of the system of nuclei. Unit volume of the substance has a total nuclear magnetic moment oriented along the external field (magnetization) of M = μ $[N(+\mu)-N(-\mu)]$. Using expression (1.1) we may write

$$\frac{N(+\mu)-N(-\mu)}{N(+\mu)+N(-\mu)} = \frac{e^{\frac{2\mu H}{kT}}-1}{1+e^{\frac{2\mu H}{kT}}}.$$ (2.1)

As usually $2\mu H/kT \ll 1$, then $\exp(2\mu H/kT) \approx 1 + 2\mu H/kT$. By substituting this value in expression (2.1) and replacing $N(+\mu)-N(-\mu)$ by M/μ and $N(+\mu)+N(-\mu)$ by the total number of nuclei in unit volume of the substance N_0, we obtain

$$M = \frac{\mu^2 N_0 H}{kT}.$$ (3.1)

In a more general case, if the nuclear spin equals I,

$$M = \frac{\mu^2 N_0 (I+1) H}{3kTI}.$$ (4.1)

The proportionality coefficient between the values M and H is called the static magnetic susceptibility of the nuclei of the substance X_0 and then

$$M = X_0 H,$$ (5.1)

where

$$X_0 = \frac{\mu^2 N_0 (I+1)}{3kTI}.$$ (6.1)

A change in magnetization indicates that some of the nuclei have changed their orientation relative to the direction of the external field. The reorientation of a nucleus is accompanied by a change in its energy and therefore reorientation can occur only in the presence of some object with which the nucleus can exchange energy. Such objects may be adjacent nuclei, ions, free radicals, or a radiofrequency coil with a high Q-factor tuned to the frequency of precession of the nuclei in the external field.

The rate of change of magnetization of nuclei M in a magnetic field H_0 is determined by the expression

$$\frac{dM}{dt} = \frac{X_0 H_0 - M}{T_1},$$ (7.1)

where T_1 is the relaxation time. If the nuclei are completely isolated, the latter is infinitely great. The more a nucleus reacts with its surroundings, the smaller the relaxation time. For pure water $T_1 \approx 3.6$ sec, for pure ethanol $T_1 \approx 5$ sec, and for pure benzene $T_1 \approx 19$ sec. In the presence of air the relaxation time of water falls to 2.5 sec and of benzine, to 5 sec.

If at the initial time t = 0 the magnetization of the nuclei M is M_1, integration of expression (7.1) gives the following relation of M to time t:

$$M = M_1 + (X_0 H_0 - M_1)(1 - e^{-\frac{t}{T_1}}).$$ (8.1)

This relation is given in Fig. 1.1 for the case $M_1 < X_0 H_0$. The graph shows that the magnetization is practically equal to the equilibrium value after a time of the order of 3 T_1. Thus, if the substance is in a strong magnetic field H_0 for a time much greater than T_1, the equilibrium magnetization of the nuclei M = $X_0 H_0$. When the magnetic field strength is reduced to H_0', the magnetization starts to change at a rate determined by expression (7.1). During a short time interval $\Delta t \ll T_1$, the magnetization will decrease by

$$\Delta M = \frac{dM}{dt} \Delta t = \frac{X_0 H_0' - M}{T_1} \Delta t.$$ (9.1)

As $X_0 H_0' \ll M$, from expression (9.1) we have

$$\frac{\Delta M}{M} = -\frac{\Delta t}{T_1}.$$ (10.1)

Thus, during $\Delta t \approx 0.1$ T_1, the nuclear magnetization will decrease by only 10%, i.e., in a weak magnetic field H_0' the nuclei will have a magnetization value considerably greater than equilibrium for quite a long time.

Pound was the first to discover the nonequilibrium polarization of nuclei by a strong magnetic field when studying the resonance of Li in a very pure LiF monocrystal [50]. The crystal was placed in a magnetic field H_0 = 6400 oe for a time greater than T_1 = 300 sec. The crystal was then transferred to a weak magnetic field and the magnetization of the nuclei fell with a relaxation time $T_1 \approx 15$ sec.

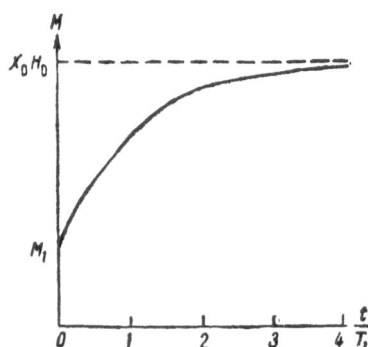

Fig. 1.1. Theoretical relation of the magnetization of the nuclei M to the time the substance remains in a magnetic field with a strength H_0.

The nuclei may also be oriented by "dynamic polarization" [51-54]. This method is based on the same interactions which lead to rapid relaxation of nuclei in the presence of paramagnetic particles and free radicals when the energy absorbed or emitted by the nucleus during its reorientation in the external magnetic field is emitted or absorbed by unpaired electrons adjacent to the nucleus. This effect may be observed when this interaction is considerable, i.e., when the presence of electrons or paramagnetic particles considerably decreases the relaxation time of the nuclei. The rapid reorientation of the magnetic moments of electrons under the influence of a radiofrequency or ultrasonic field of appropriate frequency produces either positive or negative polarization of the nuclei. If the magnetic moments of the electrons are reoriented in the external magnetic field, i.e., when electronic transitions between Zeeman levels are excited, the maximum theoretical polarization coefficient K of nuclei in metals will equal the ratio of the gyromagnetic ratios of the electron and the nucleus. For example, K = 1690 for lithium and 2400 for sodium. In solutions of paramagnetic salts and free radicals the maximum attainable polarization coefficient equals half this value. For example, in the case of protons this method can produce magnetization which is 330 times as great as the equilibrium value.

Garver and Slichter [55] were the first to observe this effect experimentally in metallic lithium and sodium dispersed in oil and in a solution of sodium in ammonia. By saturating the electron resonance line, in a field corresponding to a nuclear resonance frequency of 50 kc they obtained an increase in the nuclear polarization by a factor of 110 for lithium, 10 for sodium, and 150 for ammonia protons. Bennet and Torry [56] observed a negative Overhauser effect in a solution of sodium and naphthalene in 1,2-dimethoxyethane. When the electron resonance in a field of 17.8 oe was saturated, the polarization of the solvent protons became negative and its absolute value exceeded that of the equilibrium polarization by a factor of 65. A greater effect could be obtained in a weak magnetic field by exciting electronic transitions between the levels of the hyperfine structure. For example, theoretically it is possible to obtain a polarization coefficient of water protons equal to 3880 using a solution of Fremy's salt (potassium nitrosodisulfonate) in a field of 0.5 oe.

Abraham, Cambrisson, and Solomon [57] were the first to achieve dynamic polarization by this method although the polarization coefficient was only 100. A polarization coefficient of 1000 was obtained in a later work [58].

By using a polarizing magnetic field of 10,000 oe it is possible to obtain in a field of 0.5 oe a polarization coefficient which exceeds by a factor of 20 the maximum value of K determined in a solution of Fremy's salt and exceeds by a factor of 60 the maximum value K = 330 obtained with stable free radicals. It should be noted that the achievement of a dynamic polarization coefficient close to the maximum value involves a considerable decrease in the relaxation time of the nuclei. This produces broadening of the nuclear resonance line and rapid decrease in the magnetization of nuclei as the liquid passes from the polarizer to the detector. We may conclude from the above that at present it is more advantageous to use polarization of the nuclei by a strong magnetic field in a nuclear resonance flow gauge.

2.1. Polarization of a Flowing Liquid by a Strong Magnetic Field

For polarizing a flowing liquid by a magnetic field, the liquid is passed through a certain volume in the interpolar space of a magnet, in a solenoid, or in Helmholtz coils. The liquid has to remain in this volume for a time comparable with the relaxation time T_1, then it flows through a connecting tube, lying in a weaker magnetic field, and through a nuclear resonance detector. The polarizing system is designed to produce magnetization of the nuclei M in the liquid flowing through the detector. Let us find the relation of M to the parameters of the polarizer and the connecting tube.

In order that a liquid flowing through a certain volume shall remain in that volume as long as possible with the same flow rate, it is necessary to ensure that it has the same velocity at any point in the cross section of the volume. For this purpose special distributing devices are placed in the polarizers. If this condition is fulfilled, the time the liquid remains in the polarizing field

$$t = \frac{v_p}{q},$$

where v_p is the geometric volume of the polarizer and q the liquid flow rate. With a liquid flow rate which is nonuniform across the cross section of the polarizer, $t < v_p/q$. Let us assume that the strength of the polarizing field H_p is the same throughout the polarizer and that the liquid flows into the latter with a magnetization M_1. Then by replacing t by v_p/q and H_0 by H_p in expression (8.1) we obtain the magnetization M_2 in the liquid flowing out of the polarizer,

$$M_2 = X_0 H_p \left(1 - e^{-\frac{v_p}{qT_1}}\right) + M_1 e^{-\frac{v_p}{qT_1}}. \tag{11.1}$$

Two conclusions may be drawn from an examination of this expression. Firstly, the condition $v_p/qT_1 \gg 1$, must be fulfilled to obtain maximum polarization of the liquid flowing from the polarizer. As the graph in Fig. 1. shows, the value of M_2 differs from the maximum value $X_0 H_0$ by less than 5% if $v_p/qT_1 > 3$. Secondly, when the condition for maximum polarization is fulfilled, the magnetization of the nuclei M_1 in the liquid flowing into the polarizer contributes nothing to the value of M_2. The polarized liquid passes into the detector through a connecting tube lying in a magnetic field with a strength H_t. If we neglect the nonuniformity of the velocity of the liquid across the cross section of the tube, then knowing the distribution of the magnetic field strength along the tube, we can determine the magnetization of the nuclei M of the liquid emerging from the tube by integrating an expression similar to (7.1). In practice the effect of the field H_t is small and therefore the calculations are sufficiently accurate if we assume that H_t has a similar value throughout the whole volume of the connecting tube v_t. The value M may be determined from expression (8.1) by replacing t by v_t/q, H_0 by H_t, and M_1 by M_2 in it:

$$M = X_0 H_t \left(1 - e^{-\frac{v_t}{qT_1}}\right) + M_2 e^{-\frac{v_t}{qT_1}}. \tag{12.1}$$

The condition $v_t/qT_1 \ll 1$ must be fulfilled in order to avoid considerable demagnetization of the liquid in the connecting tube. In this case $\exp(-v_t/qT_1) \approx 1 - v_t/qT_1$ and expression (12.1) may be simplified:

$$M = X_0 H_t \frac{v_t}{qT_1} + M_2 \left(1 - \frac{v_t}{qT_1}\right). \tag{13.1}$$

This simplification introduces an error of less than 1% if $v_t/qT_1 \le 0.14$ and less than 10% if $v_t/qT_1 \le 0.34$. By substituting M_2 from formula (11.1) in formula (12.1) we obtain a general expression for the magnetization of the nuclei M at the outlet of the polarizing apparatus:

$$M = X_0 H_p (1 - e^{-\frac{v_p}{qT_1}}) e^{-\frac{v_t}{qT_1}} + X_0 H_t (1 - e^{-\frac{v_t}{qT_1}}) + M_1 e^{-\frac{v_p + v_t}{qT_1}}. \tag{14.1}$$

When the condition $v_p/qT_1 \gg 1$, is fulfilled, the contribution of M_1 may be neglected in expression (14.1) and then

$$M = X_0 H_p \left(1 - e^{-\frac{v_p}{qT_1}}\right) e^{-\frac{v_t}{qT_1}} + X_0 H_t \left(1 - e^{-\frac{v_t}{qT_1}}\right). \tag{15.1}$$

If the connecting tube is placed in a weak magnetic field, i.e.,

$$H_t \left(1 - e^{-\frac{v_t}{qT_1}}\right) \ll H_p e^{-\frac{v_t}{qT_1}},$$

14

then the expression for M is simplified further:

$$M = X_0 H_p \left(1 - e^{-\frac{v_p}{qT_1}}\right) e^{-\frac{v_t}{qT_1}}. \tag{16.1}$$

It is readily seen that both conditions may be reduced to the condition that the magnetization M in the absence of a polarizing field ($H_p = 0$) must be less than the magnetization in the presence of this field, i.e.,

$$M(H_p = 0) \ll M(H_p \neq 0).$$

With a sufficiently small volume of the connecting tube when $v_t/qT_1 \ll 0.1$ expression (16.1) may be simplified in the same manner as expression (13.1) to give

$$M = X_0 H_p \left(1 - e^{-\frac{v_p}{qT_1}}\right)\left(1 - \frac{v_t}{qT_1}\right). \tag{17.1}$$

3.1. Experimental Investigation

To check the expressions obtained it is necessary to determine the magnetization of the nuclei M when the various parameters of the polarizing apparatus are changed.

In Chapter 3 we will show that if no significant magnetization of the nuclei is produced in the liquid during its passage through a nuclear resonance flow detector, i.e., practically all the magnetization of the liquid in the detector has been introduced into it outside, then the nuclear resonance signal amplitude must be proportional to the magnetization of the nuclei M in the liquid entering the detector. This condition may be fulfilled by placing the detector in a weak magnetic field with a strength H_0 satisfying the condition $X_0 H_0 \ll M$, or by using a detector with a small volume so that there is a very rapid change of the liquid and the time the nuclei are in the detector is much less than their relaxation time T_1. In both cases the relation of the nuclear resonance signal amplitude to the parameters of the polarizing apparatus should correspond to the expressions found in the previous paragraph.

For checking this, experimental relation of the absorption signal amplitude to the strength of the polarizing field was determined [12, 13]. The working substance was water, which flowed from a water supply into a volume $v_p = 300$ cc in the gap of a polarizing electromagnet, the strength of whose field was measured with a Weber gauge with an accuracy of 1% and varied over the range of 500 to 2000 oe. The connecting tube was 200 cm in length and had a cross section of 2 cm². The coil of the detector circuit was 5 cm in length and was wound in two layers with thin wire on a glass tube with a cross section of 2 cm². An autodyne detector was used for detecting the signal [59]. The liquid flow rate was about 300 cc/sec. The field was modulated with a frequency of 15 Hz and the signal-to-noise ratio in a field of 20 oe when $H_p = 2000$ oe was about 20. The experimental results are given in Fig. 2.1. The experimental points corresponding to expression (16.1) fall on a straight line.

The same experimental apparatus was used to check the relation of the signal amplitude to v_t and v_p at a constant flow rate, but $H_p = 2000$ oe was produced with an electromagnet with an interpolar space of 5000 cc so that a volume $v_p = 2$ liters could be used and a liquid flow rate $q = 2$ liter/sec produced with a centrifugal pump.

The dimensions of the variable volume designed in the form of a cylinder with a piston (Fig. 3.1) necessitated a high flow rate. When the piston was pushed in completely, the space within the cylinder was 400 cc and when it was pulled out, the space was increased by a definite amount Δv_t. A system of funnels and grids was used to produce a uniform flow rate of the liquid across the cross section of the volume [60].

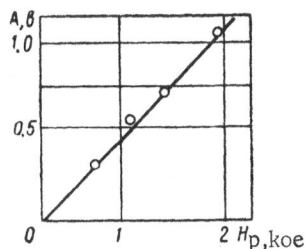

Fig. 2.1. Experimental relation of the nuclear resonance signal amplitude in a flow detector to the strength of the polarizing field.

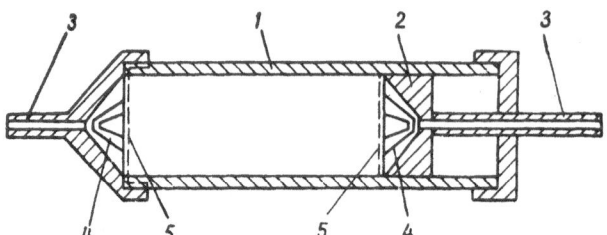

Fig. 3.1. Cross section of the variable volume with uniform distribution of the liquid flow rate across the cross
section. 1) Cylinder; 2) piston; 3) connecting pieces
for the entry and exit of the liquid; 4) distribution funnels; 5) equalizing grids.

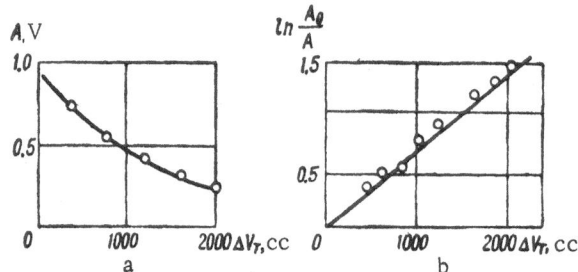

Fig. 4.1. Experimental relation of the nuclear resonance
signal amplitude to the volume of the connecting tube.
a) Linear scale; b) logarithmic scale (A_0 is the signal
amplitude at $\Delta v_t = 0$).

In order to obtain the relation of A to the volume of the connecting tube, a cylinder in a weak field was connected into the tube between v_p and the working volume of the detector. Fig. 4a.1 gives the relation obtained. Figure 4b.1 gives the same experimental points in different coordinates. From expression (16.1) the theoretical relation will be

$$\ln \frac{A_0}{A} = \frac{\Delta v_{\text{т}}}{q T_1} \,.$$

The experimental points (see Fig. 4b.1) corresponding to this formula fall on a straight line. The slope of the lines increases when paramagnetic ions are added to the water and this corresponds to a decrease in the longitudinal relaxation time.

For obtaining the relation of the signal amplitude to the volume of liquid in the magnetizing field, the vessel of variable volume was placed in a strong field (Fig. 5.1). The strength of this field and the liquid flow rate were kept constant. As the polarizing field was nonuniform it was necessary to average the process of magnetization of the liquid in separate elementary volumes within which the field could be considered as uni-form in order to find the theoretical relation of the signal amplitude to v_p. As a result of this difficulty, the curve in Fig. 5.1 was not interpreted in detail, though its general form corresponded to the theoretical relation (16.1) obtained with the assumption that the polarizing field is uniform. Thus, the results of the experimental check confirmed the accuracy of the theoretical expressions.

4.1. Practical Designs of Polarizers

The polarizing apparatus must fulfill two basic requirements: it should ensure a high magnetization of the nuclei in the liquid emerging from it and it should produce little distortion of the main magnetic field in which

16

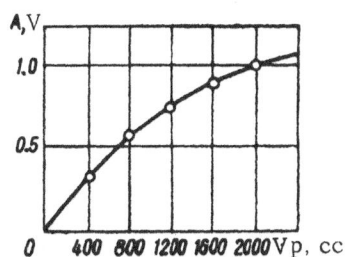

Fig. 5.1. Experimental relation
of the nuclear resonance signal
amplitude in a flow detector to
the volume of the vessel in the
polarizing field.

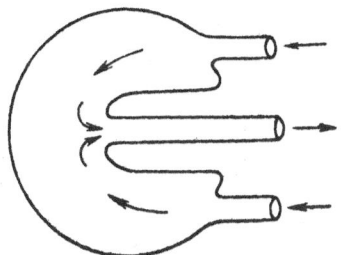

Fig. 6.1. Plan of polarizer
from [7].

the nuclear resonance detector is placed. These requirements are contradictory as to increase the magnetization of the nuclei it is necessary to
increase the strength of the polarizing field and the volume of the magnet
and decrease the size of the connecting tube, bringing the polarizing field
closer to the main field. At the same time, to weaken the effect of the
polarizing and the main fields it is necessary to decrease the strength of
the polarizing field and the size of the magnet and to place the polarizing
field further away from the main field.

With a polarizing field of high strength, iron-encased magnets are
used to produce it in order to reduce its effect on the main field as far as
possible [11, 12, 18, 19, 27, 46]. For example, in one of the instruments
[27] the magnet used had the form of a cylinder with a diameter of 52
and a height of 48 cm; the interpolar space had a diameter of 143 and a
height of 24 mm and when $H_p \approx 20{,}000$ oe, the stray field did not exceed
$3 \cdot 10^{-4}$ oe at a distance of 2m.

In the observation of the nuclear induction signal in the earth's
field, the flowing liquid was polarized in a vessel in a system of Helmholtz coils, which was designed so that the strength of the polarizing field
fell very rapidly with an increase in the distance from the coils [13]. With
a field in the polarizing volume of several hundred oersteds the field at
a distance of 1 m did not exceed a few microoersteds [61]. With a given
volume of the interpolar space and a given liquid flow rate, to obtain
the maximum residence time of the nuclei in the polarizing field it is
necessary to ensure that there is a uniform flow of liquid across the cross
section of the polarizing volume. Various types of distributing devices
are used for this purpose. The plan of a polarizer used in early work [7]
is given in Fig. 6.1. As the authors mentioned, this design could not ensure the maximum residence time of the nuclei in the polarizing volume
t = v/q (v is the volume of the polarizer and q the liquid flow rate). In some apparatuses the polarizing vessel
consisted of a glass coil [8] or a metal chamber containing a spiral which controlled the liquid flow [11, 12,
18, 19, 46]. The designs of polarizing and demagnetizing vessels with various systems for preventing stagnation
of the liquid are shown in Fig. 7.1.

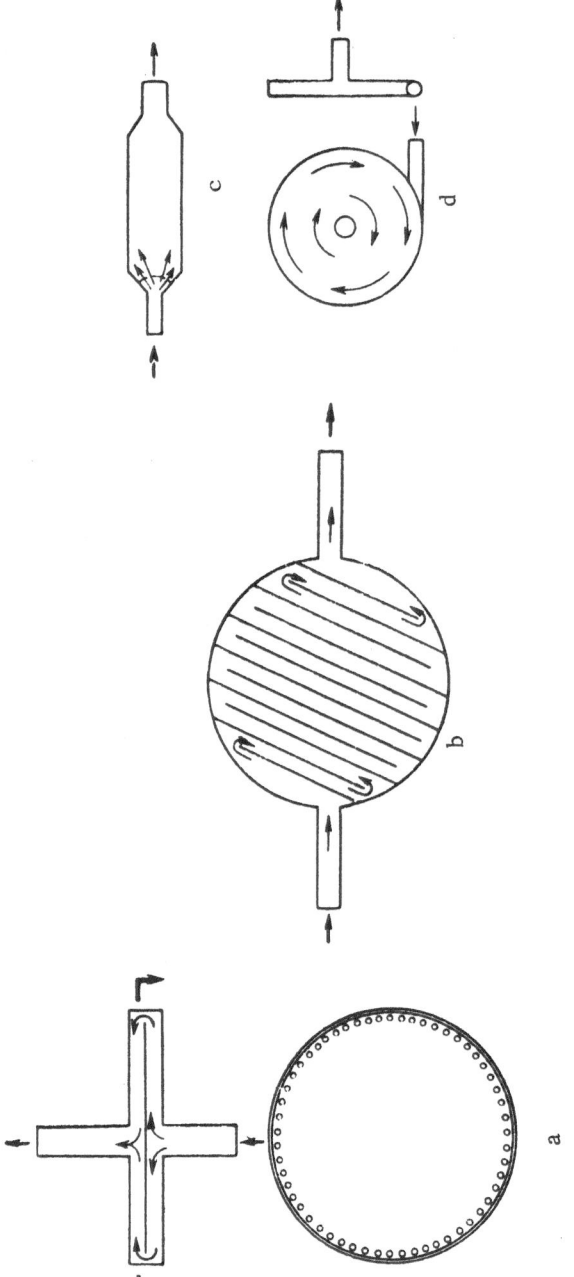

Fig. 7.1. Some polarizer designs: a) polarizer for a magnet with cylindrical poles with holes in the ends; b) polarizer for a magnet with cylindrical poles without holes; c) cylindrical polarizer; d) polarizer with centrifugal distribution of the liquid.

CHAPTER 2

INVERSION OF MAGNETIZATION OF NUCLEI IN A FLOWING LIQUID

1.2. Inversion of Magnetization of Nuclei by Rapid Reversal of the External Magnetic Field

A substance which has spent a long time in a constant magnetic field has an equilibrium magnetization of the nuclei parallel to the field. With a slow change in the direction of the field the magnetization of the nuclei rotates together with it. If the rotation of the magnetic field occurs in a time t which is much less than the precession period of the nuclei

$$t \ll \frac{2\pi}{\gamma H_0},$$

where γ is the gyromagnetic ratio of the nuclei and H_0 is the field strength, the magnetization is unable to turn and deviates from the direction of the field. This effect was first observed by Purcell and Pound [63]. They observed the nuclear resonance of lithium in an LiF crystal in a magnetic field with a strength of 6400 oe at room temperature. The crystal, which was initially in a strong field, was rapidly placed in a small solenoid whose axis was parallel to a field with a strength of about 100 oe, produced by a small permanent magnet. The solenoid was connected in series with a resistor of 500 ohm and through it was discharged a condenser with a capacitance of 2 µf with a potential on the plates of 8 kV. Thereupon, for a very short time (about 2 µsec) the field in the coil was reversed relative to the field of 100 oe and then slowly returned to the original value with a time constant of the order of 1 msec. The Larmor precession period in a field of 100 oe is 6 µsec, i.e., the magnetization of the nuclei could not follow the rapid reversal of the field and became negative. The return of the field to the initial value occurred over a time which is much greater than the precession period and therefore the magnetization remained antiparallel to the field. After this the crystal was rapidly returned to the strong magnetic field and a nuclear resonance signal was observed. This signal was positive, i.e., the system of nuclei radiated energy into the radiofrequency coil. The change in the signal occurred with a relaxation time of the lithium nuclei T_1 = 300 sec. It fell to zero and then became negative and its amplitude increased to the original equilibrium value.

Reorientation of the magnetization of the nuclei by rapid rotation of the external field was used by Packard and Varian [64] in measuring the earth's field by the free precession method. In this case, a considerably stronger polarizing field was superposed on the earth's field perpendicular to it. When equilibrium magnetization of the nuclei had been established in this field it was rapidly removed. The magnetization was unable to rotate and remained perpendicular to the earth's field. As a result of precession, an emf was induced in the receiving coil and its frequency made it possible to determine the field strength. In a flow detector the reorientation of the magnetization in the way described above may be achieved by several methods.

In the work of Hrynkiewicz et al. [65, 66] the water was polarized in flowing through a strong magnetic field (Fig. 1.2), then it flowed out of the magnet gap through a tube into the weak

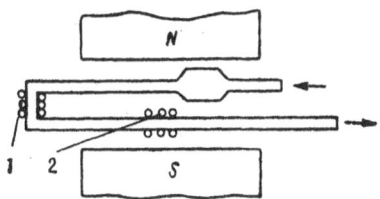

Fig. 1.2. Plan of apparatus for inversion of the nuclear magnetization in a flowing liquid by rapid reversal of the external magnetic field: 1) Solenoid for reversal of the field; 2) coil of the NMR detector circuit.

stray field, passed inside a solenoid 1, and again entered the interpolar space, where it flowed through a nuclear magnetic resonance detector 2. When a current pulse was passed through the solenoid, the field inside it changed its direction in about one microsecond and then the proton resonance signal changed in magnitude for a certain time. This experiment was similar to that carried out by Purcell and Pound. With a rapid change in the direction of the field, the magnetization of the nuclei of the liquid, which at the given moment was inside the solenoid, was directed at an angle to the external field. In this work the authors obtained the relation of the amplitude of the modified proton resonance signal to the angle Θ between the direction of the field produced by the solenoid and the direction of the stray field of the magnet at the solenoid. The signal amplitude was found to be proportional to the cosine of the angle.

As in Purcell and Pound's experiment, with a rapid change in the direction of the external field the magnetization of the nuclei in the liquid which at the given moment was inside the solenoid remained parallel to the stray field of the magnet. The strength of the solenoid field considerably exceeded the strength of the stray field and therefore the direction of the total field lay at an angle Θ to the direction of the stray field. Thus, after the discharge of the condensers the magnetization of the nuclei in the liquid in the solenoid was at an angle Θ to the direction of the external field. The component of the magnetization in the direction of the external field, which was proportional to the cosine of Θ fell very rapidly due to the nonuniformity of the external field as the precessing magnetic moments of individual nuclei were out-of-phase. During the flow of the liquid from the solenoid to the nuclear resonance detector the transverse component disappeared practically completely and therefore liquid with magnetization of the nuclei proportional to the cosine of Θ flowed into the detector. As already mentioned in Ch.1, with a sufficiently rapid flow of liquid through the detector and a long relaxation time, the signal amplitude is proportional to the magnetization of the nuclei carried by the liquid. For this reason, for a time equal to v/q (v is the volume inside the solenoid and q is the liquid flow rate) the proton resonance signal amplitude was proportional to the cosine of Θ.

By rapid reversal of the external field it is also possible to achieve continuous inversion of the magnetization of the nuclei in a flowing liquid. For this purpose the magnetic field must have a topography such that in a certain section of the trajectory of the motion of the liquid it changes sign abruptly. When the liquid passes through this section the magnetization of the nuclei is unable to reverse with the field and liquid with negative magnetization flows from this section.

In the work of A. I. Zhernovoi and G. D. Latyshev [67] the magnetization reversal device examined was joined into the connecting tube. A decrease or reorientation of the magnetization of the nuclei was recorded as a decrease or a change in the polarity of the nuclear resonance signal. If the change in direction of the field occurs in a section of the trajectory of the nuclei of length l, then in a system of coordinates moving together with the nuclei with a velocity W, the rotation of the field occurs in a time l/W. For the magnetization of the nuclei to be unable to follow the rotating vector of the external field, this time must be much less than the precession period of the nuclei at the moment of rotation $2\pi/\gamma H$, where H is the mean field strength over the section l. The distance is determined by the geometric dimensions of the device providing the required topography of the magnetic field, which are related to the diameter of the tube d. Assuming that $l > d$, we find

$$W \gg \frac{\gamma H d}{2\pi} .$$

(1,2)

This condition shows that inversion of the magnetization must be observed in very weak fields. For example, when W = 100 cm/sec and d $\approx$ 0.4 cm, from expression (1.2) it follows that H < 0.08 oe. Therefore, to obtain inversion of the magnetization of the nuclei, the section of the tube passed through a space with walls of soft iron which screened the external magnetic fields and by moving this section inside the space, a position was found in which negative magnetization of the nuclei was observed in the emerging liquid.

In one experiment the moveable section of tube consisted of rubber tubing with an internal diameter of 4 mm coiled into a ring with a diameter of 2 cm. The magnetic shield was an iron cylinder with a diameter of 10 and a height of 15 cm with walls 5 mm thick. The cylinder was placed vertically or horizontally in the stray field of the polarizing magnet. The ends were closed with iron caps and through an opening in one of the ends passed the tubes from the moveable section. The position of the section inside the cylinder at which the magnetization fell was found by moving these tubes. Reversal of magnetization was observed with more accurate adjustment. If this effect could not be obtained, it was sufficient to change the position of the cylinder relative to the polarizing magnet so that it appeared.

In another experiment two cylindrical iron rods with channels along the axes were used. They were put onto a section of the tube and placed in a magnetic shield. With various relative positions of the cylinders and shield, in the flowing liquid there was a decrease or rotation through 180° of the magnetization of the nuclei. The effect was caused by the fact that the weak residual magnetization of the cylinders was in opposite directions in them and therefore when the liquid flowed from one cylinder to the other there was abrupt reversal of the field.

By the method described it is possible to obtain continuously a liquid with negative polarization by simple means. No additional energy sources are required for this as the magnetic moments of the nuclei are reoriented at the expense of the energy of the flowing liquid.

The method may also be used for intermittent reversal of the magnetization of a liquid. For this purpose it is sufficient to wind a few turns of wire on the section of the tube in which reorientation of the magnetization of the nuclei occurs. When a current of a few milliamperes is passed through the turns the magnetic reversal effect disappears because of the distortion of the topography of the magnetic field. Intermittent passage of a current leads to intermittent reversal of the magnetization of the nuclei.

The condition for inversion of the magnetization of the nuclei (1.2) may be made less rigid by using repeated reversal of the external field. The deviations of the magnetization of the nuclei occurring in separate reversals of the field must be summed and therefore it is necessary to fulfill the phasing condition, i.e., to guarantee that during the time between reversals of the field the angle of precession of the magnetization of the nuclei is an odd multiple of π. Then for a deviation of the magnetization of the nuclei from the direction of the external field by an angle of π it is sufficient that during one reversal there is a deviation by an angle of π/n, where n is the number of reversals. It is simplest to carry out a double reversal of the field (n = 2) in the form of a change in the direction of the field for a time Δt. With fulfillment of the phasing condition $\gamma H_0 \Delta t = (2k + 1)\pi$, (k is an integer and H_0 is the field strength in the time interval Δt), complete inversion of the magnetization of the nuclei will occur if each change in direction of the field produces nutation of the magnetization through an angle of $\pi/2$. For this purpose it is sufficient for the reversal time of the external field to be of the same order as the precession period of the nuclei, i.e., as a condition for inversion we can use the inequality (1.2), changing the sign $\gg$ for the sign $\geq$. Thus, the use of repeated reversal of the external field makes it possible to effect inversion of the magnetization of the nuclei with less screening of the external field and a lower velocity of the nuclei. In practice, for inversion of the magnetization of the nuclei by double reversal of the field, onto a section of the tube lying in a weak magnetic field was placed a miniature Helmholtz coil, producing a magnetic field opposite in direction to the external field. With an increase in the current supplying the coil, the magnetization of the nuclei M in the liquid emerging from it changed polarity in accordance with the phasing condition.

This effect may also be used for measuring the flow rate of a liquid W by establishing the current at which the magnetization of the nuclei in the emerging liquid equals zero as a result of the two reversals of the field. This corresponds to fulfillment of the phasing condition $\Delta H_0 l_0 /W = (2k + 1) \pi/2$, where l_0 is the diameter of the coil and ΔH_0 is the total strength of the external field and the field produced by the coil. With a very small change in W the phasing condition no longer holds and this produces a change in the magnetization of the nuclei.

2.2. Rotation of Magnetization of Nuclei by an Oscillating Magnetic Field

Introduction. In a substance in a steady magnetic field H_0 subjected to a pulsed weak magnetic field in a direction across the steady magnetic field, oscillating with a frequency equal to the precession frequency of the nuclei, the magnetization of the nuclei deviates from the direction of the field H_0 by some angle Θ (nutation angle), which, if we neglect the relaxation processes is determined by the expression

$$\Theta = \gamma H_1 \Delta t,$$

where γ is the gyromagnetic ratio of the nuclei, H_1 is half the strength of the oscillating field, and Δt is the time of action of the oscillating field. By selecting the values of H_1 and Δt it is possible to turn the magnetization of the nuclei through any angle to the field H_0. This effect is essentially the basis of the whole phenomenon of nuclear magnetic resonance (NMR). It is used directly in pulse methods of observing a signal [68-73].

The rotation of the magnetization of nuclei in a flowing liquid by an oscillating field was described for the first time by Sherman [4, 5], who used Bloch's induction method. The liquid flows through a transmitting coil, in which, under the action of an oscillating magnetic field, the magnetization of the nuclei is rotated through an angle of $\pi/2$ relative to the direction of the external field, and then the liquid flows into another coil where the precessing magnetic moment of the nuclei gives an induction signal, which is recorded in the usual way.

The same effect was used by F. I. Skripov for observing the signal of free precession in the earth's field [13, 61]. The liquid was polarized in the field of a solenoid. Onto the tube from the polarizing field to the detector was wound a radiofrequency coil, where, as in Sherman's experiment, the magnetization of the nuclei was rotated through an angle of $\pi/2$. The liquid entering the nuclear resonance detector was magnetized at right angles to the external field, as was necessary for observing the free precession signal. For increasing the accuracy of the measurement of the earth's field, in Skripov's apparatus the alternating potential from the output of the receiver of the free precession signal was fed to the rotating coil. The system then worked as an oscillator with a frequency close to the nuclear precession frequency.

A. I. Zhernovoi was the first to observe the periodic nutation effect [18, 19]. A liquid polarized preliminarily was passed through a radiofrequency coil, which lay in a weak uniform field, and then entered a nuclear resonance detector, where it gave a signal proportional to the magnetization of the nuclei. When a resonance oscillating field was produced in the coil, the nuclear resonance signal changed in amplitude and sign, giving information on the movement of the magnetization on nutation. By increasing the strength of the oscillating field it was possible to observe up to 10 nutation periods. As many methods of using nuclear resonance in a flowing liquid in practice have been developed on the basis of the nutation effect, we should make a more detailed examination of the characteristics of this phenomenon, which has received little attention in the literature [18-21].

Theory. An oscillating magnetic field is equivalent to the sum of two fields of half the amplitude rotating in opposite directions in a plane normal to the vector of the external steady field. The component rotating in the same direction as the precessing magnetic moments of the nuclei participates in the nuclear resonance effect. The second component produces a shift of the resonance frequency, i.e., the Bloch-Siegert shift [74]. The interaction of a rotating magnetic field with the magnetization of the nuclei is most conveniently examined in a system of coordinates rotating together with the radiofrequency field.

Let the x axis lie along the vector of the rotating field and the z axis along the vector of the external field. At the initial moment the magnetization of the nuclei M is directed along the z axis. In precessing under the influence of the external field, it suffers a deviation from the z axis. The rate of increase of the nutation angle equals γH_1, where γ is the gyromagnetic ratio of the nuclei and H_1 is the strength of the rotating field. Having deviated from the z axis, the magnetization begins to precess under the influence of the external field with a strength H_0 with an angular velocity of $\gamma H_0 - \omega$ is the angular velocity of the rotation of the axes of the coordinates (the frequency of the oscillating field). Let us denote the projections of the magnetization on the axes of the coordinates by M_x, M_y, and M_z and examine the rules governing their changes. It is readily shown that M_z changes only as a result of precession of the magnetization under the influence of the field H_1. If γ is positive the precession produces a deviation of the magnetization in a clockwise direction and the rate of change of M_z is given by

$$\frac{dM_z}{dt} = M_y \gamma H_1. \tag{2.2}$$

M_x changes only as a result of precession of the magnetization under the influence of the field H_0. Its rate of change

$$\frac{dM_x}{dt} = -(\gamma H_0 - \omega) M_y. \tag{3.2}$$

As a result of the two motions, M_y changes at a rate

$$\frac{dM_y}{dt} = -M_z \gamma H_1 + M_x (\gamma H_0 - \omega). \tag{4.2}$$

The equations obtained are valid over a time interval which is much less than the relaxation times T_1 and T_2. The longitudinal spin-lattice relaxation produces a change in M_z at a rate

$$\frac{dM_z}{dt} = \frac{X_0 H_0 - M_z}{T_1}, \tag{5.2}$$

where X_0 is the static nuclear magnetic susceptibility. Transverse relaxation produces a change in M_x and M_y at a rate

$$\left. \begin{aligned} \frac{dM_x}{dt} &= -\frac{M_x}{T_2}, \\ \frac{dM_y}{dt} &= -\frac{M_y}{T_2}. \end{aligned} \right\} \tag{6.2}$$

Thus, the change in the magnetization of nuclei lying in a constant magnetic field of strength H_0 under the influence of a magnetic field of strength H_1, rotating at a frequency ω in a plane normal to the field H_0, is described by a system of three equations:

$$\left. \begin{aligned} \frac{dM_z}{dt} &= M_y \gamma H_1 + \frac{X_0 H_0 - M_z}{T_1}, \\ \frac{dM_x}{dt} &= -(\gamma H_0 - \omega) M_y - \frac{M_x}{T_2}, \\ \frac{dM_y}{dt} &= -M_z \gamma H_1 + M_x (\gamma H_0 - \omega) - \frac{M_y}{T_2}. \end{aligned} \right\} \tag{7.2}$$

The solution of such a system for conditions of sharp resonance tuning, i.e., when the frequency of the oscillating field ω equals the precession frequency of the nuclei γH_0, is given in appendix 1. In this system the effective relaxation times of the longitudinal and transverse components of the magnetization of the nuclei are denoted by T_{1n} and T_{2n}. It will be shown below that these "nutation" relaxation times are normally less than the true times.

In a flow detector the effect of the oscillating field on the nucleus begins at the moment when the nucleus enters the coil of the detector and ends when the nucleus passes out of the coil. If v_n is the volume of liquid inside the coil and q is the liquid flow rate, the action of the oscillating field on a nucleus continues for a time v_n/q. Thus, in a flow detector the oscillating magnetic field acts on the nuclei in the form of a pulse of duration v_n/q. The form of the pulse depends on the topography of the magnetic field along the route of the liquid. For simplicity, let us use a rectangular topography of the oscillating magnetic field in the coil, i.e., the strength of the field is constant along the route of the liquid in the coil and falls sharply to zero at the entrance and exit. Then the expressions for the components of the magnetization of the nuclei in the liquid emerging from the nuclear resonance detector $M_{x\ ex}$, $M_{y\ ex}$, and $M_{z\ ex}$ may be obtained by substituting in the corresponding expressions in appendix 1, $t = v_n/q$.

In the subsequent flow of the liquid, the component M_z changes with a relaxation time T_1 according to the law

$$M_z = M_{z\,ex}\ e^{-\frac{t}{T_1}} + X_0 H (1 - e^{-\frac{t}{T_1}}). \tag{8.2}$$

If the strength of the field H in which the liquid flows after emerging from the detector is low enough $X_0 H \ll M_z$, then

$$M_z = M_{z\,ex}\ e^{-\frac{t}{T_1}}. \tag{9.2}$$

The components M_x and M_y fall with an effective relaxation time T_2^*:

$$M_x = M_{x\,ex}\ e^{-\frac{t}{T_2^*}}, \quad M_y = M_{y\,ex}\ e^{-\frac{t}{T_2^*}}.\} \tag{10.2}$$

In a nonuniform external field the magnetic moments of the nuclei precessing in a field with slightly different strengths get out of phase and this leads to a rapid fall in M_x and M_y, i.e., to a low value of T_2^*. A time t after the emergence of the liquid from the detector, the magnetization M is determined by the expression

$$M = \sqrt{ M_{z\,ex}^2\ e^{-\frac{2t}{T_1}} + (M_{x\,ex}^2 + M_{y\,ex}^2)\, e^{-\frac{2t}{T_2^*}} }. \tag{11.2}$$

As $T_2^* \ll T_1$, when $t \gg 2T_2^*$, the second term in the expression for M is much less than the first and may be neglected so that then

$$M = M_{z\,ex}\ e^{-\frac{t}{T_1}}, \tag{12.2}$$

i.e., the magnetization of the nuclei of the liquid after the nuclear resonance detector is proportional to $M_{z\,ex}$. The expression for the projection $M_{z\,ex}$ has the form:

$$M_{z\,ex} = \left\{ \left(X_0 H_p - X_0 H_0 Z_n \frac{T_{1n}}{T_1} \right) \left[\frac{e^{b\frac{v_n}{q}} + e^{-b\frac{v_n}{q}}}{2} + \frac{(e^{b\frac{v_n}{q}} - e^{-b\frac{v_n}{q}})(T_{1n} - T_{2n})}{4b T_{1n}\, T_{2n}} \right] + \right.$$

$$\left. + X_0 H_0 (1 - Z_n) \frac{e^{b\frac{v_n}{q}} - e^{-b\frac{v_n}{q}}}{2b T_1} \right\} e^{-\frac{v_n}{2q}\left(\frac{1}{T_{1n}} + \frac{1}{T_{2n}}\right)} + X_0 H_0 Z_n \frac{T_{1n}}{T_1} \ldots, \tag{13.2}$$

where

$$Z_n = \frac{1}{1 + \gamma^2 H_1^2 T_{1n} T_{2n}}; \quad b = \sqrt{ \left(\frac{1}{2T_{2n}} - \frac{1}{2T_{1n}} \right)^2 - \gamma^2 H_1^2 }.$$

Experimental Investigation. As has already been mentioned, the nutation of the magnetization of nuclei by a resonance oscillating field in a flowing liquid has been investigated experimentally by means of an apparatus with polarization of the flowing liquid by a strong magnetic field and two nuclear resonance detectors in series (Fig. 2.2).

For the polarization we used an iron-clad electromagnet with a space between the poles v_p = 400 cc and a field strength H_p = 5000 oe. The nutation detector 1 consisted of a single-layer radiofrequency coil 10 mm long, wound on a tube with an internal diameter of 4.5 mm. The geometric volume of the detector v_n = 0.16 cc. The absorption detector 2 consisted of a two-layer coil 10 mm long, wound with PÉ-0.15 wire onto a bulge in the glass tube with an internal diameter of 18 mm. The coil of the nutation detector was connected to the output of ZG-12 audiofrequency oscillator and the coil of the absorption detector to the circuit of an autodyne nuclear resonance detector 4. A change in the magnetization of the nuclei in the nutation detector was recorded as a change in the signal amplitude at the output of the nuclear resonance detector.

If the strength of the magnetic field in the absorption detector H_a was much lower than the strength of the polarizing field H_p, the nuclear resonance signal amplitude was proportional to the magnetization of the nuclei in the liquid entering the detector. In the experimental apparatus used, H_a = 30 oe, while H_p = 5000 oe, i.e., this condition applied. The time for movement of the liquid t from the nutation detector to the absorption detector depends on the volume of the tube v_t connecting the detectors and the liquid flow rate q:

$$t = \frac{v_t}{q} .$$

From expression (12.2) the magnetization M of the nuclei in the liquid emerging from the absorption detector is proportional to the projection of the magnetization of the nuclei in the liquid emerging from the nutation detector and depends on the time of flow of the liquid between detectors:

$$M = M_{z\,ex}\ e^{-\frac{v_t}{qT_1}}. \tag{14.2}$$

The amplitude of the nuclear resonance signal A is proportional to this value of M.

In the absence of the conditions for nuclear resonance in the nutation detector, in the latter there is only a fall in the magnetization of the nuclei as a result of the natural spin-lattice relaxation, when

$$M_{z\,ex} = M_p\, e^{-\frac{v_t}{qT_1}}, \tag{15.2}$$

$$M = M_p\, e^{-\frac{v_n}{qT_1}} e^{-\frac{v_t}{qT_1}}, \tag{16.2}$$

where M_p is the magnetization of the nuclei in the liquid flowing into the nutation detector. The nuclear resonance signal amplitude A_0 is proportional to this value of M in the absence of nutation. Let us designate as the amplitude of the nutation signal A_n the relative change in the amplitude of the nuclear resonance signal as a result of the nutation effect:

$$A_n = \frac{A - A_0}{A_0} . \tag{17.2}$$

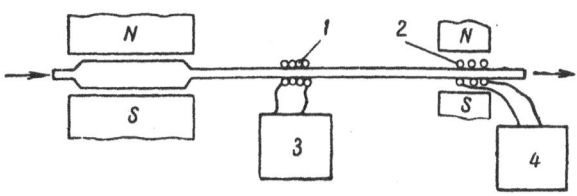

Fig. 2.2. Plan of apparatus for investigating the nutation effect; 1) nutation detector; 2) absorption detector; 3) oscillator; 4) nuclear resonance detector.

By replacing A and A_0 by the values of the nuclear magnetization proportional to them from expressions (14.2) and (16.2) we obtain the expression

$$A_n = \frac{M_{zex} - M_p e^{-\frac{v_n}{qT_1}}}{M_p e^{-\frac{v_n}{qT_1}}} = -\left(1 - \frac{M_{zex}}{M_n} e^{+\frac{v_n}{qT_1}}\right).$$

(18.2)

In the experimental apparatus the strength of the field in the nutation detector $H_n = 5$ oe. It is much less than the strength of the polarizing field H_p, which determines the value of the magnetization M_p, and therefore we can neglect terms containing H_0 in expression (13.2) for M_{zex} and

$$A_n = -A\left[1 - e^{-K}\left(\frac{e^{\sqrt{K_1^2-\Theta^2}} + e^{-\sqrt{K_1^2-\Theta^2}}}{2} + \frac{e^{\sqrt{K_1^2-\Theta^2}} - e^{-\sqrt{K_1^2-\Theta^2}}}{2\sqrt{1-\frac{\Theta^2}{K_1^2}}}\right)\right]$$

(19.2)

where A is the amplitude of the absorption signal in the absence of nutation;

$$K_1 = \frac{v_n}{2q}\left(\frac{1}{T_{2n}} - \frac{1}{T_{1n}}\right),$$

(20.2)

$$\Theta = \frac{v_n}{q}\gamma H_{1n},$$

(21.2)

$$K = \frac{v_n}{q}\left(\frac{1}{2T_{2n}} + \frac{1}{2T_{1n}} - \frac{1}{T_1}\right).$$

(22.2)

Expression (19.2) may be checked with the experimental relation of the nutation signal amplitude to the strength of the oscillating field in the nutation detector at several values of q (Fig. 3.2). The liquid flow rates are given on the curves.

From an examination of the theoretical expression it is readily shown that a periodic dependence of the nutation signal amplitude on Θ must be observed when $K_1 < \Theta$. In this case, expression (19.2) assumes the form

$$A_n = -A\left[1 - e^{-K}\cos\sqrt{\Theta^2-K_1^2} + \frac{\sin\sqrt{\Theta^2-K_1^2}}{\sqrt{\frac{\Theta^2}{K_1^2}-1}}\right].$$

(23.2)

With an increase in Θ, the amplitudes of the nutation signal vary about the value A, having extrema (minima or maxima) where the following condition holds:

$$\sqrt{\Theta^2-K_1^2} = n\pi.$$

(24.2)

The first extremum of the nutation signal must be when $\Theta = \sqrt{K_1^2 + \pi^2}$, the second when $\Theta = \sqrt{K_1^2 + 4\pi^2}$, etc. The amplitudes of the extrema satisfy the condition

$$A_{n\,extr} = -A[1 - e^{-h}(-1)^n].$$

(25.2)

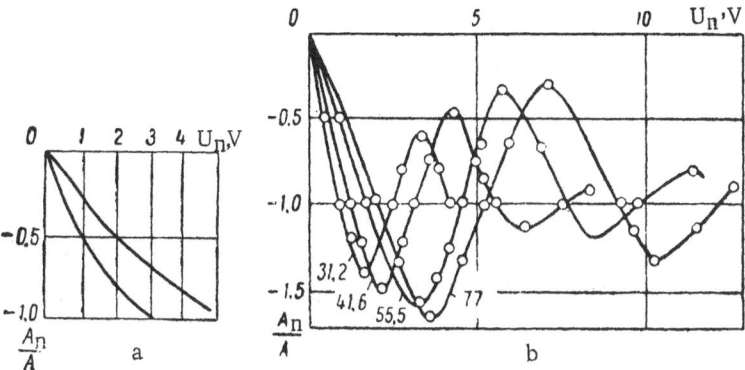

Fig. 3.2. Experimental relation of the nutation signal amplitude to
the strength of the resonance oscillating field and the liquid flow rate:
a) nonuniform external magnetic field in detector directed at right
angles to the flow; b) nonuniform external field in detector directed
along the flow.

The amplitude of the variations in A_n with an increase in Θ forms the value Ae^{-K}, i.e., a periodic dependence
of K_n on Θ will be observed only when $K < 4$. If $K > 4$ or $K_1 > \Theta$, then the change in A_n with an increase in Θ
has an aperiodic character. In particular, if $K_1 \gg 1$, then expression (19.2) has the form

$$A_n = -A\,[1 - e^{-K+\sqrt{K_1^2 - \Theta^2}}]. \tag{26.2}$$

The theoretical relations of the nutation signal amplitude to the strength of the oscillating field in the
nutation detector are given in Fig. 4.2.

The curves in Fig. 4a.2 were calculated by means of expression (26.2) for K = 4, 10, and 25 and the curves
in Fig. 4b.2, by means of expression (19.2) for $0.5 \leq K \leq 4$. As K_1 was unknown, in both cases it was assumed
that $K_1 = K$. Along the abscissa axis we plotted the ratio Θ/K, which is proportional to the strength of the oscillating field, as follows from expressions (21.2) and (22.2). A comparison of the theoretical and experimental
curves shows that, firstly, the experimental points obtained at high flow rates correspond to a smaller value of
K; this confirms qualitatively that K is proportional to 1/q. Secondly, the appearance of a nonuniform transverse field in the volume of the detector increases K. This means that transverse nonuniformity of the external
field reduces the relaxation time.

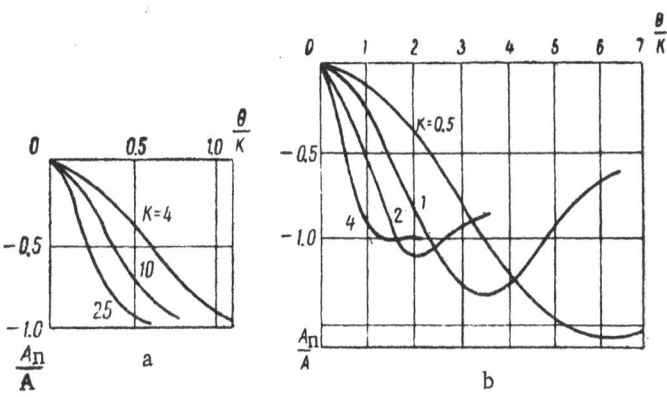

Fig. 4.2. Theoretical relation of the nutation signal amplitude to the
strength of the oscillating field and the liquid flow rate: a) k ≥4; b) k ≤4.

For a more detailed check on the accuracy of expression (22.2) it is possible to use the relation

$$\ln \left| \frac{A}{A - A_{n\,extr}} \right| = K \qquad (27.2)$$

where $\varphi = \pi$, $2\pi\ldots$, which follows from condition (25.2). Here $A_{n\,extr}$ is the extreme (maximum or minimum) value of the nutation signal amplitude and A is the amplitude of the absorption signal in the absence of nutation.

The experimental relation of K to the liquid flow rate is given in Fig. 5.2. The value of $\ln | A/(A - A_{n\,extr})|$ was determined from the experimental graph in Fig. 3b.2. Along the abscissa axis we plotted the reciprocal of the liquid flow rate at which we observed the corresponding extreme nutation signal amplitude $A_{n\,extr}$. In Fig. 5a.2 the experimental points denoted by crosses refer to first extrema and those denoted by circles, to second extrema. The linear relation obtained corresponds to the theoretical formula (25.2) with the condition that T_{1n} and T_{2n} are constant. The theoretical slope of the line drawn through the experimental points relative to the ordinate axis when $T_{2n} = T_{1n}$ equals T_{2n}/v_n. By determining this slope from the graph in Fig. 5a.2 we find that $T_{2n}/v_n = 0.035$, whence we find that when $v_n = 0.16$ cc, $T_{2n} = 0.006$ sec. If we assume that $T_{2n} \ll T_{1n}$, from the same data we find that $T_{2n} = 0.003$ sec.

A similar graph constructed from the amplitude of the third nutation extremum is given in Fig. 5b.2. In this case the slope of the line is three times that in the graph in Fig. 5a.2, i.e., the transverse relaxation time T_{2n} is less by a factor of three.

The experimental data given in Fig. 3b.2 makes it possible to check the validity of expression (21.2) and estimate K_1. From expression (23.2) it follows that the amplitude of the nutation signal has extrema at $\varphi = \pi$ and $\varphi = 2\pi$ and when the following condition holds:

$$\tan \varphi = \frac{\varphi}{K_1}, \qquad (28.2)$$

the amplitude of the nutation signal $A_n = -A$. If $K_1 = 0$, then condition (28.2) holds when $\varphi = \pi/2$ and $3\pi/2$. Having the relation of A_n to the potential on the nutation coil U_n, it is possible to construct the relation of φ to U_n. It is assumed that U_n corresponding to the appearance of the first extremum $\varphi = \pi$, at the appearance of the second extremum $\varphi = 2\pi$, and when $A_n = -A$, then on the assumption that $K_1 \ll \pi/2$, $\varphi = \pi/2$ and $3\pi/2$.

Thus, from the data on the graph in Fig. 3.2 we constructed the relation of φ to U_n, which is given in Fig. 6.2. Figure 7.2 gives the same relation in different coordinates.

The lines obtained show that φ is proportional to H_{1n}/q. This is true of $K_1 \ll \pi/2$ as then $\varphi = \Theta = \gamma H_{1n} v_n/q$. If $K_1 > 0.23$, then condition (28.2) holds when φ is less than $\pi/2$ by more than 10%; if $K_1 > 0.4$, then it holds when φ is less than $\pi/2$ by more than 20%. The graph (see Fig. 6.2) shows that the ordinates of the corresponding points deviate from $\pi/2$ by no more than 10%, i.e., $K_1 < 0.23$.

The curves in Fig. 3b.2 correspond to the values $K = 0.9 - 0.4$. In the case when $K = 0.9$, $K_1/K < 0.25$. From expressions (20.2) and (22.2) when $T_1 \gg T_{1n}$

$$\frac{K_1}{K} = \frac{T_{1h} - T_{2h}}{T_{1h} + T_{2h}} < 0.25,$$

whence $T_{1n} > T_{2n} > 0.6 T_{1n}$. Consequently, with the nonuniformity of the external field directed mainly along the flow of the liquid in the detector, the effective longitudinal and transverse relaxation times are similar in magnitude. As was established, $T_{2n} \approx 0.10^{-3}$ sec, i.e., the effective relaxation times are considerably less than the true relaxation times $T_2 \approx T_1 \approx 2$ sec.

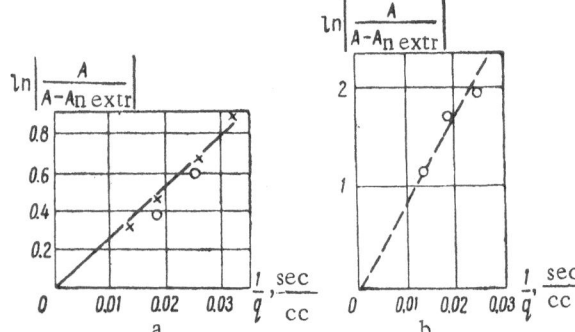

Fig. 5.2. Experimental relation of K to the liquid flow rate: a) first and second nutation extrema; b) third nutation extremum.

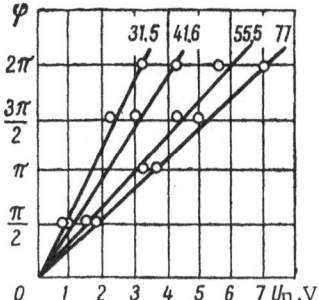

Fig. 6.2. Experimental relation of the nutation angle of the nuclear magnetization to the strength of the oscillating field and the liquid flow rate.

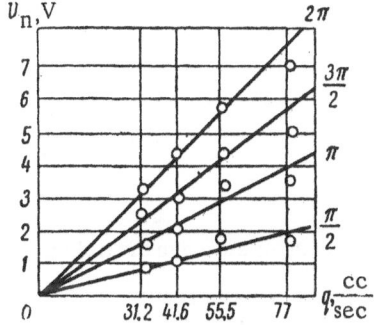

Fig. 7.2. Experimental relation of the strength of the oscillating field to the liquid flow rate and the nutation angle of the nuclear magnetization.

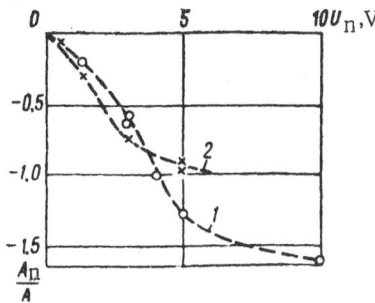

Fig. 8.2. Experimental relation of the nutation signal amplitude to the strength of the oscillating field in the nutation detector: 1) with longitudinal nonuniformity of the field; 2) with transverse nonuniformity of the field.

The experimental relations of A_n to H_{1n} for two positions of the nutation detector are given in Fig. 8.2. In one case the liquid flowed with the gradient of the external field and in the other, perpendicular to it.

Curve 1 corresponds to longitudinal nonuniformity of the field and curve 2 to transverse nonuniformity.

The theoretical relation of A_n/A to Θ is given in Fig. 9.2. Curve 1 was constructed for the case when $K = K_1 = 0.5$. With a change in K_1, the shape of the curve changes only slightly when $\Theta \ll 1$, i.e., the position of the point at which $A_n = -A$ and the position of the extremum does not change. The value $K = 0.5$ was chosen so that the amplitudes of the extrema on the theoretical and experimental curves were equal. Curve 2 was constructed for the case when $K = K_1 = 4$. Smaller values of K do not correspond to the smooth form of the experimental curve 2 in Fig. 8.2 as from Fig. 4b.2 it follows that curves for $K < 4$ have appreciable extrema.

An examination of Figs. 8.2 and 9.2 shows that the theoretical curve with $K = K_1 = 4$ lies more to the right than the corresponding experimental curve (curves with $K = K_1 > 4$ lie even further to the right). The theoretical curve is displaced to the left with a decrease in K_1 relative to K. Let us find the relation of K and K_1 at which the positions of the corresponding theoretical and experimental curves coincide. For this it is necessary to ensure that with $\Theta = \pi/2 = 1.57$, when the theoretical curve 1 passes through -1, the theoretical curve 2 should pass through the point with the ordinate -0.86, as occurs on the experimental graph (see Fig. 8.2).

At high values of K and K_1 [expression (19.2)], A_n assumes the form

$$\left.\begin{aligned} \frac{A_n}{A} &= e^{-K+\sqrt{K_1^2-\Theta^2}} - 1 \\[2mm] \ln\frac{A_n+A}{A} &= -K + \sqrt{K_1^2-\Theta^2}. \end{aligned}\right\} \quad (29.2)$$

or

By substituting $\Theta = 1.57$ and $(A_n + A)/A = 0.14$, we obtain the relation of K_1 and K:

$$\left.\begin{aligned} -1.96 &= -K + \sqrt{K_1^2-2.5} \\[2mm] K_1 &= \sqrt{2.5+(K-1.96)^2}. \end{aligned}\right\} \quad (30.2)$$

or

Estimation of K from expression (22.2) gives the value $K = 100-200$. When $K = 100$, $K_1 = 98$, i.e., $K_1/K = 0.98$, while when $K = 200$, $K_1 = 198$, i.e., $K_1/K = 0.99$. If we allow for the fact that the longitudinal nonuniformity of the field decreases somewhat the effective working volume of the detector and this displaces the theoretical curve 1 to the right (see Fig. 9.2), bringing it nearer to curve 2, the difference in K_1 and K will be still less. In any case, with transverse nonuniformity of the field $K > K_1 > 0.98$ K, or $1 \geq K_1/K \geq 0.98$. By substituting the values of K_1 and K from expressions (20.2) and (22.2) we find $1 > (T_{1n}-T_{2n})/(T_{1n}+T_{2n}) > 0.98$, or $T_{2n} < 0.01\ T_{1n}$. This means that the transverse nonuniformity of the field preferentially reduces to the transverse relaxation time T_{2n}.

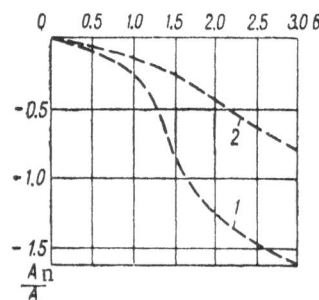

Fig. 9.2. Theoretical relation of the nutation signal amplitude to the nutation angle: 1) $K = K_1 = 0.5$; 2) $K = K_1 = 4$.

Conclusions. The experimental investigation of the amplitude of the nutation signal showed that expression (19.2) accurately reflects the processes occurring in a nutation detector. This expression was derived on the assumption that there are two relaxation times of the longitudinal component of the nuclear magnetization: T_1 and T_{1n}. The natural spin-lattice relaxation time T_1 is related to the natural increase in the magnetization of the nuclei in the nutation detector under the influence of the steady external magnetic field H_n. The rate of increase in the magnetization component M_z produced by this process is given by

$$\frac{dM_z}{dt} = \frac{X_0 H_n}{T_1} . \tag{31.2}$$

The relaxation time T_{1n} is related to the decrease in the component M_z in the nutation detector. This process is analogous to the decrease in the components M_x and M_y and the rate of decrease in M_z produced by it is determined by an expression analogous to expression (6.2):

$$\frac{dM_z}{dt} = - \frac{M_z}{T_{1n}} . \tag{32.2}$$

The total rate of change of M_z

$$\frac{dM_z}{dt} = \frac{X_0 H_n}{T_1} - \frac{M_z}{T_{1n}} . \tag{33.2}$$

When $T_{1n} = T_1$, expression (33.2) describes the normal law of the natural change in magnetization in the field H_n.

The experimental investigation showed that in the nutation detector $T_{1n} \ll T_1$, i.e., the change in the projection M_z in the flow detector is described not by equation (5.2), which was obtained at the beginning of the section, but by another equation in which allowance is made for the effect of the relaxation time T_{1n}. This equation has the form:

$$\frac{dM_z}{dt} = \frac{X_0 H_0}{T_1} - \frac{M_z}{T_{1n}} . \tag{34.2}$$

It appeared in the system of equations solved in appendix 1. The effective relaxation time T_{1n} characterizes the decrease in the component M_z for all the reasons. It includes the contribution of the natural relaxation process and also the decrease in the magnetization of the nuclei in its nutation because of the different rates of nutation of the magnet moments of separate nuclei. The effective relaxation time T_{2n} characterizes the decrease in the components M_x and M_y. It also includes the contribution of the natural relaxation process and the dephasing of the magnetic moments of the nuclei in their nutation. Moreover, the transverse components of the magnetization decrease because of the dephasing of the magnetic moments of the nuclei processing in a magnetic field of different strengths, i.e., when the external magnetic field is nonuniform.

As was established, with a low transverse gradient, $T_{1n} \approx T_{2n} = 10^{-3}$ sec, i.e., the contribution of the natural relaxation process is negligibly small. The identical decrease in the longitudinal and transverse effective relaxation times of the magnetization of the nuclei, averaged over the cross section of the flow in the nutation detector, may be explained by the effect of the nonuniformity of the strength of the oscillating field and the liquid flow rates across the cross section of the flow.

The angle of nutation Θ under the influence of a resonance oscillating field with an amplitude $2H_1$ increases at a rate $d\Theta/dt = \gamma H_{1n}$. If H_{1n} varies over the cross section of the flow by ΔH_{1n}, then the magnetic moments of the nuclei passing through the detector in different parts of the cross section of the flow are dephased at a rate $\gamma \Delta H_{1n}$ and this produces a corresponding decrease in the mean magnetization across the cross section of the flow.

If the field H_{1n} is uniform, but the liquid flow rate falls from the axis of the flow to the walls of the tube, then at each point in the detector $d\Theta/dt = \gamma H_{1n} = $ const, and the angle of nutation of the magnetization of the nuclei, moving with a velocity W, varies with a derivative $d\Theta/dx = (d\Theta/dt)/W = \gamma H_{1n}/W$. In a system of coordinates moving with the mean velocity of the liquid W_{av}, the value $d\Theta/dt = (d\Theta/dx) W_{av} = \gamma H_{1n} W_{av}/W$, i.e., in the center of the cross section of the flow the angle of nutation increases more slowly than at the walls of the tube and this produces dephasing of the magnetic moments of the nuclei and a corresponding decrease in the mean magnetization across the cross section of the flow. With turbulent flow of the liquid, an analogous effect must produce pulsation of the velocity of separate elements of the volume of liquid.

Both of the effects examined produce the same decrease in the longitudinal and transverse components of the nuclear magnetization, i.e., the effective relaxation times T_{1n} and T_{2n} produced by them are equal. This is observed with nonuniformity of the external field directed along the flow of the liquid in the detector.

If the nonuniformity of the external field is directed across the flow in the detector, then without vigorous mixing of the liquid across the flow, the magnetic moments of the nuclei passing through the detector in different parts of the cross section will precess in different magnetic fields and this results in their dephasing. This dephasing decreases only the transverse component of the magnetization, i.e., makes a contribution to T_{2n}. Therefore in a field with transverse nonuniformity when there is no transverse mixing of the liquid, it must be that $T_{2n} \ll T_{1n}$. This was observed experimentally.

Expression (19.2) was derived on condition that the frequency of the oscillator and the frequency of the precession of all the nuclei in the detector were the same and therefore it is quite valid if the nonuniformity of the field is very slight. The nutation effect in a field with a high longitudinal gradient is examined in Section 3.2.

The nutation effect with a high transverse gradient may be described by a relation similar to expression (19.2) if there is modulation of the magnetic field, as in this case the oscillating magnetic field produces nutation of the nuclei to the same extent in all parts of the cross section of the detector.

When there is no modulation, the nuclei passing through the detector have a precession frequency which differs from the frequency of the oscillating field by some value $\Delta\omega$, which depends on the part of the cross section in which the nuclei are. When there is a frequency difference, the nutation effect is described by expression (7) of appendix 4 [or relation (36.2)]. To find the over-all effect it is necessary to determine the fraction of the nuclei which pass through the detector with the frequency difference $\Delta\omega$ and taking this into account, to integrate over the whole cross section of the detector. By using the mean value theorem, with any distribution function, it is possible to write

$$M_{z\,ex} = M_p e^{-\frac{v_n}{qT_{2n}}} \frac{\Delta \overline{H}^2 + H_1^2 \cos\gamma \frac{v_n}{q} \sqrt{H_1^2 + \Delta \overline{H}^2}}{\Delta \overline{H}^2 + H_1^2},$$

where $\Delta\overline{H}$ is some value proportional to the nonuniformity of the field in the detector. This expression describes the nutation effect in a field with transverse nonuniformity.

Width of Nutation Signal. With a certain amplitude of the oscillating field established in the nutation detector, the maximum nutation signal is observed at a frequency of this field equal to the precession frequency of the nuclei in this detector. The shift of the frequency of the oscillating field at which the amplitude of the nutation signal falls to half is the half-width of the nutation signal af half-height.

The width of the nuclear resonance signal is composed of the natural width of the line and the broadenings produced by the heterogeneity of the external field, its modulation, the finite time for passing through resonance, and the instrument effect. Let us estimate the natural width of the nutation signal.

The amplitude of the nutation signal is proportional to the change due to nutation in the projection on the direction of the external field of the magnetization of the nuclei emerging from the nutation detector, i.e., the value

$$\Delta M_z = M_{z\,ex} - M_p\,e^{-\frac{v_n}{qT_1}},\qquad(35.2)$$

where M_p is the magnetization of the nuclei entering the nutation detector. The law for the change in M_z under the influence of an oscillating field with an amplitude $2H_1$ and a frequency differing from the resonance frequency by $\Delta\omega$ is derived in appendix 4 by solving Bloch's equation in a rotating system of coordinates. It has the form

$$M_z = M_p\,e^{-\frac{t}{T}}\left(1 - \frac{1-\cos\sqrt{\gamma^2 H_{1n}^2 + \Delta\omega^2}\cdot t}{\frac{\Delta\omega^2}{\gamma^2 H_{1n}^2}+1}\right),\qquad(36.2)$$

where $T = T_{1n} = T_{2n}$ is the relaxation time of the nuclei. Nuclei passing through the nutation detector suffer the action of the oscillating field for a time $t = v_n/q$, while the amplitude of the oscillating field is established in accordance with the condition for obtaining the maximum nutation signal:

$$\gamma H_{1n}\frac{v_n}{q} = \pi$$

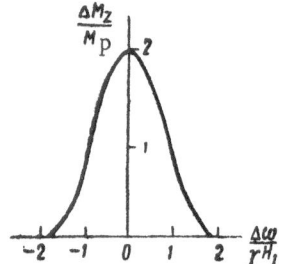

Fig. 10.2 Theoretical relation of the amplitude of the nutation signal to the difference from the resonance frequency.

(the condition for the first nutation extremum) and therefore to obtain the relation of $\Delta M_{z\,ex}$ to $\Delta\omega$ it is necessary to substitute in expression (35.2) M_z from formula (36.2) with $t = v_n/q$ and the nutation angle $\gamma H_{1n} t = \pi$. The expression obtained has the form

$$\Delta M_{z\,ex} = -M_p - \left[\frac{1-\cos\pi\sqrt{1+\frac{\Delta\omega^2}{\gamma^2 H_{1n}^2}}}{1+\frac{\Delta\omega^2}{\gamma^2 H_{1n}^2}}\,e^{-\frac{v_n}{qT}} - e^{-\frac{v_n}{qT}} + e^{-\frac{v_n}{qT_1}}\right].$$

$$(37.2)$$

A graph of this relation (with $v_n/qT \ll 1$) is given in Fig. 10.2. The graph shows that A_n, which is proportional to $-\Delta M_{z\,ex}/M_p$, decreases by a factor of two with a frequency difference $\Delta\omega = 0.8\,\gamma H_{1n}$, whence the natural width of the maximum nutation signal in a uniform field at half-height

$$\delta\omega_n = 1.6\gamma H_{1n} = \frac{5q}{v_n}.\qquad(38.2)$$

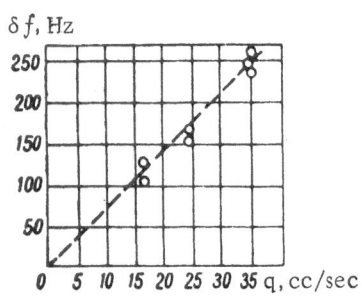

Fig. 11.2. Experimental relation of the nutation signal width to the liquid flow rate.

This expression was checked with an apparatus with two detectors (see Fig. 2.2). The nutation detector was in a uniform field with a strength of 0.5 oe, while the absorption detector was in a field with a strength of 10 oe. When there was no oscillating field in the nutation detector, a nuclear resonance signal was observed at the output of the detector circuit, which was connected to the circuit of the absorption detector and with a certain set liquid flow rate, which was measured through the time to fill a calibrated volume, the optimal conditions in the absorption detector were selected. Then an oscillating field was excited in the nutation detector, its resonance frequency adjusted for the maximum nutation effect and with this frequency, the amplitude selected which corresponded to the appearance of the maximum negative absorption signal (the first nutation extremum). Then the frequency of the field was shifted in both directions so that the amplitude of the nutation signal fell to a half (in a homogenous field this corresponds to the disappearance of the absorption signal). The total frequency shift δf equals the width of the nutation signal at the half-height.

The relation of the width at the half-height of the maximum nutation signal to the liquid flow rate in a uniform field is given in Fig. 11.2. The experimental points lay satisfactorily on a straight line, which corresponded qualitatively to the theoretical expression (38.2). To check the quantitative correspondence it is necessary to take into account the broadening of the signal due to the instrument effect and the finite time for the nuclei to pass through the detector, which, according to the conclusions in Section 2.2, for a working volume of cylindrical form is $\delta\omega_A = \pi q/4v_n$, and this added to the width determined by expression (38.2) gives the theoretical signal width at the half-height

$$\delta f = 0.925 \frac{q}{v_n} . \tag{39.2}$$

In Fig. 11.2 the broken line passing through the experimental points was drawn in accordance with this expression with $v_n = 0.13$ cc. The nutation detector had a diameter of 4.5 mm and the coil had a length of about 10 mm so that $v_n = 0.16$ cc, i.e., the theoretical estimate corresponds to the experimental value.

The external field may be considered as uniform when the nonuniformity of the field in the nutation detector $\Delta H \ll \pi q/\gamma v_n$. If this condition does not hold then the nonuniformity of the field begins to have an appreciable effect on the width of the nutation signal. The relation of the signal width to the nonuniformity of the field was determined experimentally. The nutation detector was placed in a nonuniform magnetic field and oriented so that the liquid flowed along the gradient, while the lines of force of the oscillating field were perpendicular to the external field. The frequency and strength of the oscillating field were selected so as to give the first nutation extremum. Then the nutation detector was moved along the gradient of the field until the amplitude of the nutation signal fell to a half and next it was moved in the opposite direction a distance of Δl so that the amplitude of the nutation signal again fell to half. The experiment was repeated several times with various field nonuniformities and detector geometries and in all cases the value of Δl was approximately equal to half the length of the nutation detector, i.e., the width of the nutation signal at the half-height was equal to half of the nonuniformity of the field in the volume of the nutation detector. Thus, with the nonuniformity of the field in the nutation detector $\Delta H \ll \pi q/\gamma v_n$, the width of the signal is proportional to q/v_n, while with $\Delta H \gg \pi q/\gamma v_n$, the width of the signal at the half-height equals $\Delta H/2$ and is independent of q and H_{1n}.

Multiquantum Transitions. In the investigation of the inversion of the nuclear magnetization by an oscillating field in a flowing liquid, the effect was observed not only when the frequency of the field ω was close to the frequency of precession of the nuclei in the external magnetic field H_0, but also when the factor by which it was less was a whole number, i.e., when the condition $n\omega = \gamma H_0$ (n is a whole number) held. Values of n up to 18 were found in practice.

A fundamentally similar effect might be produced by overtones from the oscillator supplying the nutation coil. This was checked by Wilking [75]. Connecting in band filters, tuned to the fundamental frequency, at the output of the oscillator hardly changed the inversion intensity at frequencies less than the Larmor frequency of the nuclei, i.e., the effect was caused by multiquantum transitions when the energy difference ΔE between the Zeeman levels of the nuclei is compensated by n photons with the frequency $\omega = \Delta E/n\hbar$. In contrast to single quantum transitions (n = 1), when, as was shown previously, the probability of transition is independent of the strength of the external field, when n > 1, this dependence is found. For the measurements we used a flow detector which is illustrated in plan in Fig. 12.2. The water was polarized in the strong magnetic field of the electromagnet 1, then flowed along a connecting tube through the nutation coil, which lay inside a solenoid 2, screened from external magnetic fields. The strength of the external field H_0 in the nutation coil

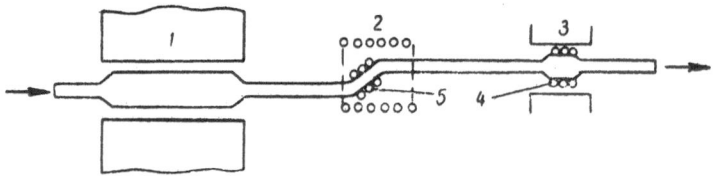

Fig. 12.2. Diagram of apparatus for investigating multiquantum transitions: 1) polarizing magnet; 2) solenoid of nutation detector; 3) magnet of absorption detector; 4) coil of absorption detector; 5) coil of nutation detector.

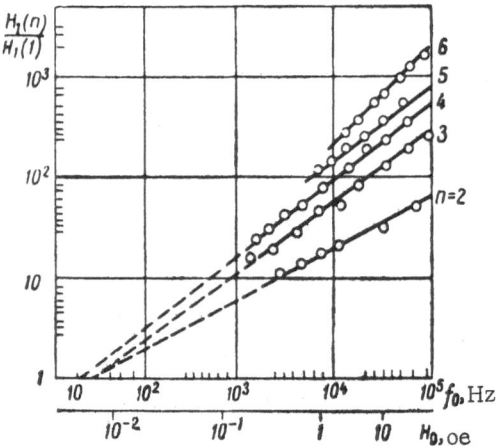

Fig. 13.2 Experimental relation of the probability of multiquantum transitions to the strength of the external magnetic field H_0 ($f_0 = \gamma H_0 / 2\pi$).

5 was regulated by changing the current in the solenoid. The water flowed from the nutation coil into the nuclear resonance detector 3, which was placed in a strong uniform magnetic field. The amplitude of the nuclear resonance signal was proportional to the magnetization of the protons in the liquid which entered the detector.

Figure 13.2 gives the experimental dependence on H_0 of the strength of the oscillating magnetic field at the n-quantum transition $H_1(n)$ required to rotate the magnetization of the protons through an angle of $\pi/2$, which was found through the disappearance of the nuclear resonance signal in the absorption detector. On a log-log scale, all the points with the same parameter n lay on straight lines, i.e., the relation may be characterized by the empirical formula

$$\ln \frac{H_1 n}{H_1(1)} = a(n) \ln \frac{f_0}{b(n)} , \qquad (40.2)$$

where a(n) is the slope of the lines and b(n) gives the points of intersection on the abscissa axis. The values of a(n) and b(n) for various values of n are given in Table 1.2.

The experimental relation given shows that the greater n and the higher the strength of the external field H_0, the greater is the strength of the oscillating magnetic field stimulating the nuclear transitions required to maintain the same rate of transitions. Thus, the probability of transitions falls rapidly with an increase in the number of quanta participating in one transition and with an increase in the strength of the magnetic field in which the nuclei lie. We should also note some peculiarities of multiquantum transitions. In contrast to single quantum transitions, where the angle of nutation of the nuclear magnetization $\theta = \gamma H_1(1)\tau$ (τ is the time for the nuclei to pass through the nutation coil), in multiquantum transitions the angle θ is not proportional to the strength of the oscillating field. With multiquantum transitions the width of the nutation signal falls with an increase in n approximately in proportion to $1/n$. The transition probability depends differently on the angle of the inclination of the nutation coil to the direction of the field H_0 for odd and even values of n.

TABLE 1.2

n	1	2	3	4	5	6
a(n)	0	0.499± ±0.003	0.668± ±0.04	0.673±0.005	0.729±0.008	0.88± ±0.01
b(n)	—	21.6±0.5	21.7±0.6	17.0±0.6	8.8±0.6	19±1

3.2. Rotation of Magnetization of Nuclei by Fast Adiabatic Passage Through Resonance

For the case of adiabatic passage through resonance, nuclei in a magnetic field of strength H, changing at a rate of dH/dt, are under the influence of an oscillating magnetic field of frequency ω_0 and amplitude H_1, directed normal to the field H. Provided that the condition $\gamma H_1^2 > dH/dt$ holds, at the moment when the strength H reaches the value $H_0 = \omega_0/\gamma$, i.e., when the frequency ω_0 equals the frequency of precession of the nuclei, there is inversion of their magnetization.

This effect was first observed in stationary samples by Bloch [76, 77] and was subsequently investigated by many scientists [78-80].

To achieve adiabatic passage through resonance in a flow detector it is necessary to pass the liquid, which has first been polarized, through a radiofrequency coil, connected to an oscillator. We will call this coil the rotation coil. It must be placed in a constant magnetic field H, whose gradient is directed at right angles to its lines of force. The axis of the coil must be directed along the gradient of the field.

If an oscillating magnetic field of strength $2H_1$ with a frequency ω_0 is excited in the coil and the strength of the external field H, selected such that in one of the sections of the coil it has the value $H_0 = \omega_0/\gamma$, with a sufficiently high strength H_1, liquid with negatively polarized nuclei will flow from the coil.

The theoretical expression describing the change under the action of the oscillating field in the magnetization of nuclei lying in an alternating field H may be obtained by solving the system of equations (7.2) and substituting in it

$$\omega = \gamma H(t).$$

We assume that $H(t) = H_0 + \Delta H(t)$, and then the system of equations assumes the form:

$$\left.\begin{aligned}
\frac{dM_z}{dt} &= -M_y\gamma H_1 + \frac{X_0 H_0 - M_z}{T_1}, \\
\frac{dM_x}{dt} &= M_y\gamma\Delta H(t) - \frac{M_x}{T_2}, \\
\frac{dM_y}{dt} &= M_z\gamma H_1 - M_x\gamma\Delta H(t) - \frac{M_y}{T_2}.
\end{aligned}\right\} \tag{41.2}$$

If the liquid passes through resonance in a short time, during which the magnetization is unable to change substantially as a result of relaxation processes, in the system of equations (41.2) we may neglect terms containing the relaxation times T_1 and T_2. Moreover, in this case the rate of change of the field H during the time of passage through resonance may be taken as independent of time and we may substitute in the equation

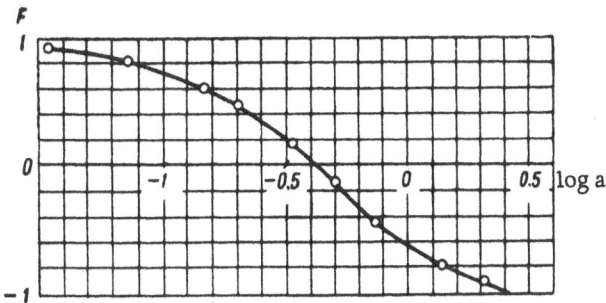

Fig. 14.2. Theoretical relation of the coefficient of rotation F of the nuclear magnetization in adiabatic passage through resonance to the value $\gamma H_1^2/(dH/dt)$.

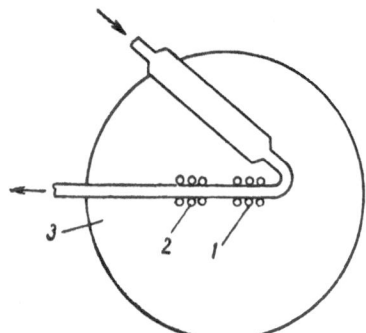

Fig. 15.2. Diagram of flow detector used to investigate fast adiabatic rotation of the magnetization of a flowing liquid: 1) rotation coil; 2) nuclear resonance detector coil; 3) magnet pole.

$\Delta H = t\,dH/dt$. Having carried out these simplifications and replaced the variable t by $\tau = \gamma H_1 t$, we obtain a new system of equations in which there is only one parameter a:

$$\left.\begin{aligned}
\frac{dM_z}{d\tau} &= -M_y, \\
\frac{dM_x}{d\tau} &= M_y\frac{\tau}{a}, \\
\frac{dM_y}{d\tau} &= M_z - M_x\frac{\tau}{a},
\end{aligned}\right\} \tag{42.2}$$

where

$$a = \frac{\gamma H_1^2}{\dfrac{dH}{dt}}. \tag{43.2}$$

This system of equations was solved for various values of a by Benoit [81] on a computer with a maximum error of 4%.

The relation of the rotation coefficient F, which equals the ratio of the value of M_z after passage through resonance to the value of M_z before, to the parameter a is given in Fig. 14.2. These data show that rotation is practicaly complete ($F \approx -1$) when a > 3, i.e., for inversion of the nuclear magnetization by fast adiabatic passage through resonance, the following condition must be fulfilled:

$$\frac{\gamma H_1^2}{\dfrac{dH}{dt}} > 3. \tag{44.2}$$

In a flow detector the rate of change of the external magnetic field acting on the nuclei is related to the liquid flow rate and the field gradient:

$$\frac{dH}{dt} = W\,\mathrm{grad}\,H.$$

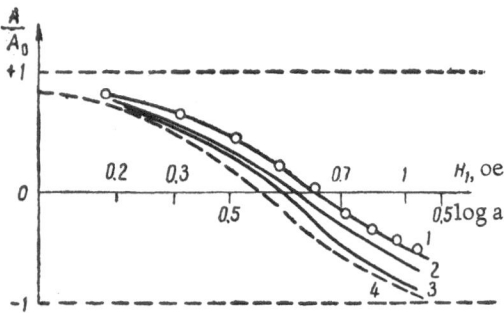

Fig. 16.2. Experimental relation of the coefficient of rotation of nuclear magnetization by adiabatic passage through resonance to the value of H_1.

By substituting this value in expression (44.2) we obtain the condition for inversion of the nuclear magnetization in a flow detector

$$\frac{\gamma H_1^2}{W\,\mathrm{grad}\,H} \geqslant 3. \tag{45.2}$$

A detector with flowing water was used for an experimental check of the theory (Fig. 15.2). A magnet with a pole diameter of 20 cm and a gap of 40 mm had a field at the center of 6900 oe, corresponding to a resonance frequency of the protons of 29.4 MHz. The water was polarized by flowing in the gap through a tube 1 cm in diameter and 20 cm long for a time which was long in comparison with the relaxation time. It then passed through a tube 2mm in diameter, through which it flowed with a velocity W = 57 cm/sec and a flow rate q = 1.8 cc/sec successively through the rotation coil and a nuclear resonance detector coil. The rotation coil was 23 mm long and had 20 turns wound on a Teflon rod 6 mm in diameter with a hole down the center through which the tube passed. The coil was connected to a carefully screened high-frequency oscillator with a power of several tens of watts, tuned to a frequency of 28.7 MHz. The amplitude of the high-frequency field in the coil could reach H_1 = 2 oe. The position of the rotation coil in the interpolar space was selected so that the mean strength of the field in it, in accordance with the chosen oscillator frequency, equaled H = 6740 oe, while the field gradient, equal to 240 oe/cm, was directed along the flow of the liquid. The rate of change of the magnetic field acting on the nuclei passing through the coil equaled 14,000 oe/sec. If it is assumed that the resonance zone extended over the region where the magnetic field differed from the resonance value by $\Delta H < 10\,H_1$, then this region had a maximum length of 1 mm, i.e., it was very narrow in comparison with the size of the coil. The field gradient could be regarded as constant within its limits. The rotation time was 0.02 sec, which is negligibly small in comparison with the natural relaxation time, as was assumed in the theoretical calculations. The frequency of the field H_1 was not critical as a change in it produced only a shift of the resonance zone along the coil. The coil of the nuclear resonance detector was connected to an autodyne detector, the amplitude of the signal from which was proportional to the magnetization of the protons in the water flowing into the detector.

In Fig. 16.2, the experimental points and curve 1 show the relation of the autodyne signal amplitude A to the strength of the oscillating field H_1 in the rotation coil.

Curves 2 and 3 show the same relation with corrections for relaxation of the nuclei in the time for the liquid to flow between the rotation coil and the nuclear resonance detector, which equaled t = 0.165 sec. The broken curve 4 shows the theoretical curve of the relation of the rotation coefficient F to a. The closest to the theoretical curve is curve 3, which was constructed from the experimental curve with the assumption that the relaxation time of the water used was 0.75 sec. The low value of the relaxation time may be explained by the impurities in the tapwater used. The small discrepancy between curve 3 and the theoretical relation may be caused by an error of 13% in the calculation of the field strength H_1 in the rotation coil. Another reason for the discrepancy may be the nonuniformity of the liquid flow rate across the cross section of the coil. Closer correspondence between the theoretical and experimental relations was obtained in analogous measurements in other work [82], in which the liquid velocity was kept equal to 100 cm/sec. With this high velocity the liquid was able to pass from the rotation coil to the coil of the nuclear resonance detector in a time which was much shorter than the relaxation time T_1 and therefore it was unnecessary to introduce corrections for demagnetization of the liquid, which had to be introduced in the previous experiment.

The gradient of the external field in the coil was 150 oe/cm and the rate of change of the strength of the external magnetic field acting on the nuclei equaled 150,000 oe/sec. To achieve the conditions for complete rotation a ≥ 3 with such a high rate of change of the field in the rotation coil, it was necessary to create an oscillating magnetic field with a strength greater than 1.5 oe. In the actual experiment the coil was supplied from a 3-watt radiofrequency oscillator and to increase H_1, it was included in a resonance circuit. The distance between the rotation and receiving coils was 5 cm. The oscillating magnetic field in the rotating coil had a frequency of 29.7 MHz and that in the NMR detector, 30 MHz, i.e., the frequency difference was only 1% and therefore careful screening was necessary to eliminate the interaction of the two coils. If the liquid was electrically conducting, the difference in the frequencies had to be increased as there was pickup through the stream which bypassed the screens.

In an apparatus for rotation of magnetization by fast passage through resonance in a weak magnetic field [83], the rotation coil was fixed in the stray field of the magnet with a strength of 10 oe. It contained 855 turns and had a length of 255 mm and a diameter of 4 mm. As the time for the liquid to pass between the polarizing field and the nuclear resonance detector was long in this experiment, the liquid used was benzene, whose relaxation time theoretically equals 19 sec. The benzene was polarized in the interpolar space of a magnet and then flowed along a tube 6 mm in diameter at a velocity of 1 meter/sec through the rotation coil into the nuclear resonance detector. The rotation coil was supplied from a low-frequency oscillator, which produced an oscillating field in it with a strength of 200 oe. The gradient of the magnetic field in the coil was directed along its axis and equaled 1.75 oe/cm. In this case the value of $\gamma H_1^2/(dH/dt)$ reached 6, i.e., the theoretical rotation coefficient equaled −1. In both experiments long coils were used so as to guarantee successful rotation over a wide range of changes in the electromagnet field.

In principle, all the designs of coil used for rotation by a resonance oscillating field may be used for rotation of nuclear magnetization by fast adiabatic passage through resonance. Rotation by an oscillating field occurs with low nonuniformity of the external field directed along or across the flow of liquid in the coil. In this case, with an increase in the strength of the resonance oscillating field H_1 there is observed a periodic change in the polarity of the magnetization of the nuclei in the emerging liquid.

Rotation by adiabatic passage through resonance occurs with a large nonuniformity of the external field directed along the flow of the liquid in the coil. In this case, with an increase in the strength of the resonance oscillating field H_1 the magnetization of the nuclei in the liquid flowing from the coil becomes negative and with a further increase in H_1 it tends to zero. The second method requires much higher strengths of the oscillating field than the first method. With considerable nonuniformity of the external field in the coil directed across the flow, rotation of the magnetization does not occur.

CHAPTER 3

NUCLEAR RESONANCE SIGNAL AMPLITUDE AND WIDTH
IN A FLOWING LIQUID

1.3 Nuclear Absorption Signal

The simplified theory examined in the introduction is usually used to explain the relation of the signal amplitude in a flow detector to the velocity of the liquid [1, 2, 84-86]. This theory corresponds satisfactorily to the experimental results obtained by observing a nuclear resonance signal in a strong magnetic field.

With an increase in the liquid velocity W, the signal amplitude increases asymptotically and tends to a constant value (see Fig. 1.I). However, when the nuclear resonance signal is observed in a weak uniform magnetic field, the signal amplitude increases with an increase in W, passes through a maximum, and then falls again. This relation is beyond the limits of the simplified theory and therefore, to explain the phenomenon of nuclear resonance in a flowing liquid we give below a more general theory, based on an examination of nutation of nuclear magnetization in a detector.

General Expression for Signal Amplitude. In the observation of a nuclear absorption signal in a flow detector, the polarized liquid moves through an oscillating magnetic field created by the coil of a radiofrequency circuit in a magnetic field of strength H_0.

As was shown in Ch. 2, if the oscillating field is directed normal to the external field and its frequency is close to the precession frequency of the nuclei $\omega_0 = \gamma H_0$, then under the influence of one of the rotating components of the oscillating field there occurs nutation of the magnetization of the nuclei from the direction of the external field. Thereupon there appear components of the magnetization directed normal to the external field. The component M_x precesses in phase with the rotating component of the oscillating field, while the component M_y precesses in quadrature with it. The appearance of transverse components of the magnetization results in the appearance in the coil of the circuit of a rotating magnetic induction with the components $B_x = 4\pi M_x$ and $B_y = 4\pi M_y$.

The rotating magnetic induction induces in the coil of the circuit an emf, whose maximum value E is determined by the maximum rate of change of the magnetic flux through the coil

$$E = -\omega_0 N S \eta \sqrt{B_x^2 + B_y^2} = -4\pi N S \eta Q \omega_0 \sqrt{M_x^2 + M_y^2}, \tag{1.3}$$

where N and S are the number of turns and the cross section of the coil, η is the degree of filling of the field of the coil by resonating nuclei, and Q is the quality factor of the circuit.

The more strict examination made in appendix 2 shows that expression (1.3) is valid only when there are no natural oscillations in the coil of the circuit, i.e., when the receiver coil is completely isolated from the coil creating the oscillating magnetic field producing the nutation of the nuclear magnetization.

If the production of the oscillating field and the reception are achieved in one coil, then the component M_y produces a change in potential in the circuit equal in magnitude to $A = 4\pi NQS \eta \omega_0 M_y$, which is called the nuclear absorption signal; the component M_x produces a change in the oscillation frequency in the circuit pro-proportional to it, which is called the dispersion signal. Normally, for convenience its nuclear signal is not detected continuously, but periodically. For this purpose, the strength of the external field or the frequency of the oscillating field is shifted periodically from the resonance value, i.e., is modulated with some frequency Ω. Thereupon the absorption signal appears as amplitude modulation of the potential in the circuit with a frequency Ω and a modulation index A, while the dispersion signal appears as frequency modulation.

In observing nuclear resonance in a stationary substance, to obtain the maximum signal amplitude it is necessary for the modulation period to exceed the relaxation time of the nuclei. In this case, during the absence of resonance conditions the nuclei are able to recover their polarization, which is saturated during the previous passage through resonance. Analogously, in a flow detector it is advantageous to select a modulation period which is considerably greater than the replacement period of the liquid in the coil, i.e.,

$$\Omega \ll \frac{q}{v},$$

where q is the liquid flow rate and v is the detector volume. Under these conditions, transient processes in the system of magnetic moments of the nuclei will not have a substantial effect on the amplitude of the nuclear resonance signal and it will be determined by the mean value of M_y through the volume of the detector, established by accurate resonance tuning.

The change in the magnetization of the nuclei occurs under the influence of the resonance oscillating field in which the nuclei are found in passing through the detector. If the external magnetic field and the oscillating field of the detector coil are uniform, the resonance process is identical for all nuclei in one section of the detector, i.e., the magnetization varies only along the flow of the liquid in the coil. In this case the law of change of the magnetization along the stream is identical to the law for its change with time, taking into account the fact that the coordinate x, measured along the direction of flow from the beginning of the detector coil, is related to time:

$$x = Wt,$$

where W is the liquid velocity in the stream.

The magnetization changes with time under the influence of the resonance oscillating field H_1 and this may be found by solving the system of equations obtained in Ch. 2, which describes the change in magnetization under the influence of a magnetic field, rotating in the direction and with the frequency of precession of the nuclei. An oscillating field is equivalent to the sum of two fields of half the amplitude, rotating in opposite directions and therefore the solution of the equation for the case of a rotating field is valid in the case of an oscillating field as the second component does not participate in the effect, producing only a slight shift in the resonance frequency [95-97].

The solution of the equations in a rotating system of coordinates on condition that there is accurate resonance tuning is given in appendix 1, where expression (20) describes the relation of the rotating component M_y of the magnetization of nuclei in a resonance oscillating field with an amplitude $2H_1$ to time. By replacing t by x/W in this expression we obtain a relation describing the change in M_y along the stream in the detector:

$$M_y = -\frac{M_p \gamma H_1}{r_1 - r_2}\left(e^{r_1 \frac{x}{W}} - e^{r_2 \frac{x}{W}}\right) - X_0 H_0 Z \gamma H_1 T_{2n}\left(1 + \frac{r_2 e^{r_1 \frac{x}{W}} - r_1 e^{r_2 \frac{x}{W}}}{r_1 - r_2}\right)\frac{T_{1n}}{T_1}, \qquad (2.3)$$

where

$$r_{1,2} = -\frac{1}{2T_{1n}} - \frac{1}{2T_{2n}} \pm \sqrt{\left(\frac{1}{2T_{2n}} - \frac{1}{2T_{1n}}\right)^2 - \gamma^2 H_1^2},$$

$$Z = \frac{1}{1 + \gamma^2 H_1^2 T_{1n} T_{2n}},$$

where M_p is the magnetization of the nuclei in the liquid flowing into the detector, X_0 the static magnetic susceptibility of the nuclei, H_1 half the amplitude of the oscillating field, T_{1n} and T_{2n} are the effective "nutation" relaxation times of the longitudinal and transverse components of the nuclear magnetization averaged over the cross section of the stream in the detector. As the mean value of the magnetization across the cross section

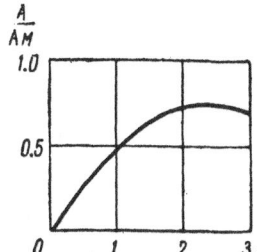

Fig. 1.3. Theoretical relation of the absorption signal
amplitude to the nutation
angle.

varies as a result of the dephasing of the magnetic moments of separate
nuclei, then $T_{1n} \ll T_1$ and $T_{2n} \ll T_2$. The signal amplitude is proportional
to the value of M_y averaged through the volume of the detector:

$$\overline{M}_y = \frac{1}{l_a} \int_0^{l_a} M_y \, dx, \tag{3.3}$$

where l_a is the length of the detector along the stream. By substituting the
value of M_y from formula (2.3) in expression (3.3) and integrating we obtain:

$$\overline{M}_y = -\frac{M_p \gamma H_1 Z T_{1n} T_{2n} W}{l_a} \left[1 - \frac{M_0 Z (T_{1n} + T_{2n})}{M_p T_1} \right] \left[1 - \right.$$

$$\left. - \left(\frac{e^{b \frac{l_a}{W}} + e^{-b \frac{l_a}{W}}}{2} + \frac{e^{b \frac{l_a}{W}} - e^{-b \frac{l_a}{W}}}{\frac{4bT_{1n}T_{2n}}{T_{1n} + T_{2n}}} \right) e^{-\frac{l_a}{2W}\left(\frac{1}{T_{1n}} + \frac{1}{T_{2n}}\right)} \right] -$$

$$- \frac{M_0 Z \gamma H_1 T_{1n} T_{2n}}{T_1} \left(1 - \frac{e^{b \frac{l_a}{W}} - e^{-b \frac{l_a}{W}}}{2bl_a} W e^{-\frac{l_a}{2W}\left(\frac{1}{T_{1n}} + \frac{1}{T_{2n}}\right)} \right), \tag{4.3}$$

where

$$b = \sqrt{\left(\frac{1}{2T_{2n}} - \frac{1}{2T_{1n}}\right)^2 - \gamma^2 H_1^2}.$$

By substituting the value of $\overline{M}_y$ found in formula (1.3), we obtain an expression for the absorption signal
amplitude:

$$A = A_M \gamma H_1 Z T_{1n} T_{2n} \frac{W}{l_a} \left\{ \left[1 - \frac{M_0 Z (T_{1n} + T_{2n})}{M_p T_1} \right] \times \right.$$

$$\times \left[1 - \left(\frac{e^{b \frac{l_a}{W}} + e^{-b \frac{l_a}{W}}}{2} + \frac{e^{b \frac{l_a}{W}} - e^{-b \frac{l_a}{W}}}{\frac{4bT_{1n}T_{2n}}{T_{1n} - T_{2n}}} \right) e^{-\frac{l_a}{2W}\left(\frac{1}{T_{1n}} + \frac{1}{T_{2n}}\right)} \right] +$$

$$+ \frac{M_0 l_a}{M_p T_1 W} \left[1 - \frac{e^{b \frac{l_a}{W}} - e^{-b \frac{l_a}{W}}}{2bl_a} W e^{-\frac{l_a}{2W}\left(\frac{1}{T_{1n}} + \frac{1}{T_{2n}}\right)} \right] \right\}, \tag{5.3}$$

where

$$M_0 = X_0 H_0, \quad A_M = -4\pi\eta N S Q \gamma H_0 M_p.$$

Expression (5.3) shows that with low strengths of the oscillating field, the signal amplitude increases with
an increase in H_1 and to obtain the maximum amplitude it is necessary for the following condition to be fulfilled:

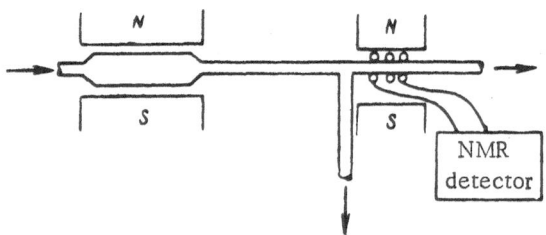

Fig. 2.3 Diagram of flow detector with preliminary polarization of the liquid with a strong field.

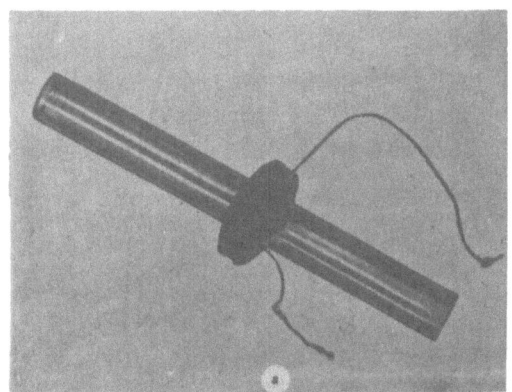

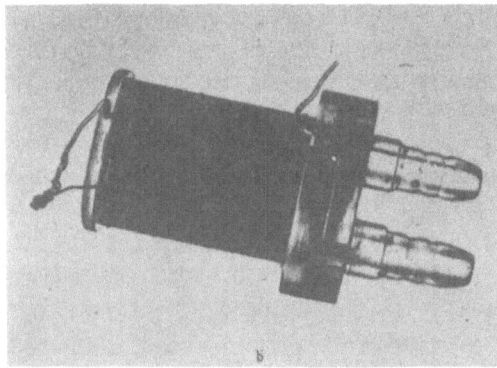

Fig. 3.3. Photographs of flow detectors: a) with a volume of 0.03 cc; b) with a volume of 15 cc.

$$\gamma H_1 \gg \frac{1}{T_{2n}} - \frac{1}{T_{1n}} ,$$
$$\frac{T_{1n} + T_{2n}}{T_1} Z \ll 1, \quad \text{i.e.,} \quad \gamma^2 H_1^2 T_1 T_{2n} \gg 1,$$
$$\frac{l_a}{2W} \left(\frac{1}{T_{1n}} + \frac{1}{T_{2n}} \right) < 0.2.$$
$$\tag{6.3}$$

In this case the signal amplitude is given by

$$A = A_M \frac{1 - \cos \Theta}{\Theta} , \tag{7.3}$$

where $\Theta = \gamma H_1 l_a / W$ is the nutation angle. This relation is shown in Fig. 1.3. The latter figure shows that the maximum signal amplitude $A_{max} = 0.7\, A_M$ is observed with a nutation angle

$$\Theta = \frac{3\pi}{4} . \tag{8.3}$$

When the conditions for maximum polarization are fulfilled, when $M_p = X_0 H_p$, the maximum signal amplitude

$$A_{max} = -2.8\pi\eta NSQ\gamma H_0 X_0 H_p. \tag{9.3}$$

To check expression (5.3) we determined the experimental relations of the absorption signal amplitude to the liquid flow rate, the strength of the oscillating field, and the angle of nutation of the magnetization of the nuclei in the detector.

Relation of the Signal Amplitude to the Liquid Flow Rate and the Detector Volume. The relation of the signal amplitude to the flow rate was investigated in the apparatus which is shown diagramatically in Fig. 2.3. The resonance of protons in water was observed. For polarization we used a permanent magnet with a volume of 100 cc between the poles and a field strength of 5000 oe. The connecting tube had a diameter of 0.3 cm and a length of 100 cm and was placed in a field with a strength of about 1 oe. The detector consisted of a coil with a universal winding 0.4 cm long, fitted on a tube with an internal diameter of 0.3 cm (Fig. 3a.3). The proton resonance was observed in a field of 30 oe by means of an autodyne detector. The liquid flow rate in the detector was varied over the range of 2-10 cc/sec. So that the signal amplitude was not affected by the flow rate dependence of the magnetization of the nuclei entering the detector, the flow rate in the polarizer and the connecting tube was kept constant, while the flow rate in the detector was varied by diverting part of the liquid through a bypass tube.

The experimental relation of the signal amplitude to the liquid flow rate q in a field with an absolute nonuniformity within the limits of the detector $\Delta H = 0.2$ oe is given in Fig. 4.3. The relation was linear.

The strength of the external field was low enough for the condition $H_0 \ll H_p$ to hold, while the appreciable nonuniformity of the field in the volume of the detector made $T_{2n} \ll T_{1n}$, as was shown in Section 2.2.

Under these conditions, from expression (5.3) we obtain the following relation of the signal amplitude to the liquid flow rate:

$$A = Bq\left[1 - \left(\frac{e^{\frac{v_a}{2qT_{2n}}\sqrt{1-4\gamma^2H_1^2T_2^2 n}}+e^{-\frac{v_a}{2qT_{2n}}\sqrt{1-4\gamma^2H_1^2T_2^2 n}}}{2}\right.\right. +$$

$$\left.\left.+\frac{e^{\frac{v_a}{2qT_{2n}}\sqrt{1-4\gamma^2H_1^2T_2^2 n}}-e^{-\frac{v_a}{2qT_{2n}}\sqrt{1-4\gamma^2H_1^2T_2^2 n}}}{2\sqrt{1-4\gamma^2H_1^2T_2^2 n}}\right)e^{-\frac{v_a}{2qT_{2n}}}\right], \qquad (10.3)$$

where B is a coefficient independent of the flow rate and v_a is the detector volume. An examination of this expression shows that if

$$4\gamma H_1 T_{2n} > 1 \text{ and } \frac{v_a}{qT_{2n}} \geqslant 4$$

or

$$4\gamma H_1 T_{2n} < 1 \text{ and } \frac{v_a}{qT_{2n}} > \frac{4+\ln\frac{1+\sqrt{1-4\gamma^2H_1^2T_2^2 n}}{\sqrt{1-4\gamma^2H_1^2T_2^2 n}}}{1-\sqrt{1-4\gamma^2H_1^2T_2^2 n}},$$

then with an error of less than 1%, $A = Bq$. With a nonuniformity in the detector of 0.2 oe, $T_{2n} = 3.7 \cdot 10^{-4}$ sec, $v_a/qT_{2n} \approx 8$, while an estimate of H_1 shows that $\gamma H_1 T_{2n} \approx 0.14$. Under these conditions, from expression (10.3) it follows that the signal amplitude depends linearly on the flow rate and this corresponds to the experimental results. Figure 4b.3 shows the relation of the signal amplitude to the flow rate q in a field with an absolute non-uniformity within the limits of the detector $\Delta H = 3.5 \cdot 10^{-4}$ oe for several values of the potential on the coil. With the same nonuniformity of the field and $T_{2n} \approx 0.2$ sec, an estimate of the strength H_1 showed that for the three different curves $\gamma H_1 T_{2n}$ has the values 5, 8, and 12. With the condition $H_0 \ll H_p$, $T_{2n} \ll T_{1n}$, and $\gamma H_1 T_{2n} > 1$, expression (5.3) has the form

$$A = A_M \frac{1-\cos\frac{v_a}{2qT_{2n}}\sqrt{4\gamma^2H_1^2T_{2n}^2-1}\frac{\sin\frac{v_a}{2qT_{2n}}\sqrt{4\gamma^2H_1^2T_{2n}^2-1}}{\sqrt{4\gamma^2H_1^2T_{2n}^2-1}}}{\gamma H_1 \frac{v_a}{qT_{2n}}}. \qquad (11.3)$$

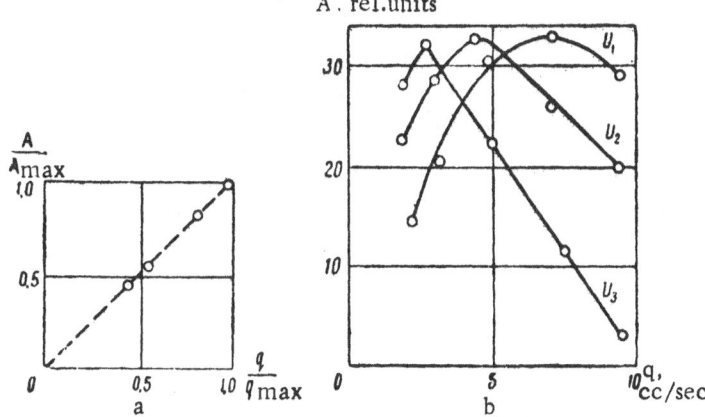

Fig. 4.3. Experimental relation of the nuclear resonance signal amplitude to the liquid flow rate a) nonuniformity of the field in the detector with $\Delta H = 0.2$ oe; b) nonuniformity of the field in the detector with $\Delta H = 3.5 \cdot 10^{-4}$ oe.

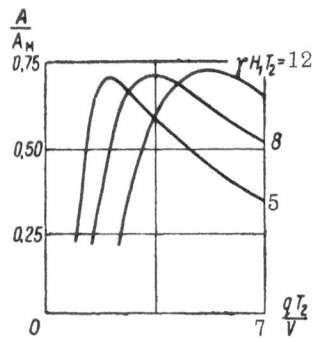

Fig. 5.3. Theoretical relation of the nuclear resonance signal amplitude to the liquid flow rate.

This relation is given in Fig. 5.3. In the same way as on the experimental curves, the signal amplitude has a maximum at some optimal flow rate with this optimal flow rate increasing with an increase in H_1.

Thus, the experimental results, which are given in Fig. 4.3, are not contrary to the theoretical expression (5.3). Let us find if this expression corresponds to the relations of the signal amplitude to the flow rate obtained in other experiments [1, 8]. In them $H_p = H_0$ and, moreover, the conditions $\gamma H_1 T_{2n} \ll 1$ and $v_a/qT_{2n} \gg 1$ hold. For example, in the apparatus of Hyrnkiewicz and Waluga [8], $T_1 = 2.3$ sec; the heterogeneity in the volume of the detector $\Delta H = 0.04$ oe or $T_{2n} = 2 \cdot 10^{-3}$ sec and $Z = 1/3$ $-1/10$, whence $\gamma H_1 T_{2n} = 0.04-0.08$, the length of the coil $l_a = 1.5$ cm, and the maximum liquid velocity $W_{max} = 40$ cm/sec, whence $v_a/q_{max}T_{2n}$ $= l_a/W_{max}T_{2n} = 19$. With these conditions holding, after some simplifications, from expression (5.3) we obtain the following value for the signal amplitude:

$$A = A_M \gamma H_1 Z T_{1n} T_{2n} \frac{W}{l_a} \left(1 - Z \frac{T_{1n}}{T_1}\right)(1 - e^{-\frac{l_a}{WT_{1n}Z}}) + A_M \gamma H_1 Z \frac{T_{2n}T_{1n}}{T_1}. \tag{12.3}$$

This expression corresponds to the experimental relations given in Figs. 1.I and 2.I. With zero liquid velocity, the signal amplitude $A = A_M \gamma H_1 Z T_{2n}$, and with an increase in velocity, it increases until the condition $l/WT_{1n}Z > 3$ holds; with a further increase in the velocity, the increase in the amplitude slows, tending to a constant value. With $W = 0$, from expression (5.3) we obtain the normal relation of the signal amplitude to the detector parameters with a stationary working material

$$A = -4\pi\eta NSQH_0^2\gamma^2 X_0 H_1 Z T_{2n}$$

(with $W = 0$, $T_{2n} \approx T_2$).

Thus, expression (5.3) accurately reflects the relation of the signal amplitude to the flow rate. From it it follows that when $\gamma^2 H_1^2 T_{1n} T_{2n} \gg 1$ and the conditions $v_a/qT_{2n} > 4$ and $H_0 v_a/H_p qT_1 \ll 1$ hold, the signal amplitude increases linearly with the liquid flow rate and is independent of its relaxation time. In actual fact, when $\gamma^2 H_1^2 T_{2n} T_{1n} \gg 1, Z = 1/\gamma^2 H_1^2 T_{1n} T_{2n}$ and from expression (5.3), $A = A_M q/\gamma H_1 v_a$. By using formula (8.3) we obtain the amplitude under the optimal conditions $A = -2.8\pi \eta QNS \gamma H_0 M_p$. With a change in the volume of the detector, its cross section S, the number of turns N, and the Q-factor change. The cross section changes in proportion to the square of the linear dimension of the detector d^2, the Q-factor changes approximately in proportion to d, while the number of turns at constant inductance is proportional to $d^{-\frac{1}{2}}$. Thus, $A \sim d^{5/2}$, i.e., the signal amplitude must increase with an increase in the linear size of the detector and this is observed experimentally.

Relation of the Signal Amplitude to the Strength of the Oscillating Field and the Angle of Nutation of the Nuclear Magnetization. The relation of the absorption signal amplitude to the strength of the oscillating field and the nutation angle of the nuclear magnetization may be obtained by detecting the signal with a bridge circuit, whose sensitivity is not affected by the amplitude of the high-frequency potential in the circuit, while in all other cases this relation is distorted by the change in the detector sensitivity.

For the experiment we used an apparatus with two detectors, through which the liquid passed successively (see Fig. 2.2) and which were placed in a uniform field $H_0 = 12$ oe. The mesh of the first detector was connected to the bridge circuit [87] of a nuclear resonance detector, while the mesh of the second was connected to an autodyne circuit [59]. The strength of the polarizing field $H_p = 10,000$ oe, the volume of the first detector $v_1 = 0.03$ cc (see Fig. 3a.3), the volume of the second detector was 15 cc (see Fig. 3b.3), and the liquid flow rate $q = 55$ cc/sec.

44

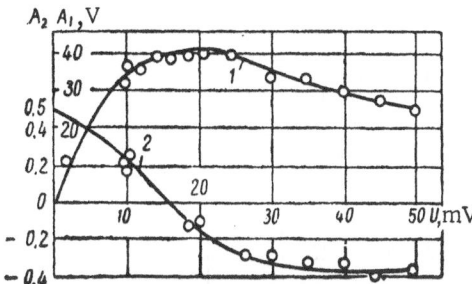

Fig. 6.3. Experimental relation of the amplitude of the nuclear resonance signal obtained in an apparatus with two flow detectors to the strength of the oscillating field in the first detector: 1) amplitude of signal in first detector; 2) amplitude of signal in second detector.

The experimental relation of the nuclear resonance signal amplitudes in the first (curve 1) and second (curve 2) detectors to the strength of the oscillating field in the first detector is given in Fig. 6.3. Curve 1 shows the experimental relation of the absorption signal amplitude A_1 to the strength of the oscillating field in the first detector. As the field H_0 is uniform, then from the results given in Section 3.2, $T_{2n} = T_{1n}$, and therefore $1/T_{2n} - 1/T_{1n} < \gamma H_1$, while the value of v_1 ensures that the conditions $v_1/qT_{2n} \ll 1$ and $\gamma^2 H_1^2 T_{1n} T_{2n} \gg 1$ hold. In this case the theoretical relation of A_1 to H_1 is described by expression (7.3) and is shown in Fig. 1.3. It may be shown that the theoretical and experimental curves correspond to each other.

Theoretically, A_1 has a maximum at $\Theta_{opt} = 2.36$. The angle of nutation Θ_{opt} at which A_1 has a maximum on the experimental curve may be determined from the amplitude of the absorption signal in the second detector (curve 2, Fig. 6.3). As has been shown (Section 2.2), in a uniform field the amplitude of this signal first becomes zero at $\Theta = 1.57$. The value of H_1 is proportional to the potential on the mesh U, i.e., the angle Θ is proportional to U. Knowing that $\Theta = 1.57$ corresponds to U = 15 mV, it is possible to determine Θ_{opt} corresponding to the maximum absorption signal. In this case U = 21 mV, i.e., $\Theta_{opt} = 2.2$. The deviation from the theoretical value is less than 7%.

The relation of the absorption signal to the oscillation voltage with an autodyne circuit for detection is given in Fig. 7.3. The maximum is much sharper than on the theoretical relation. Estimation of the optimal nutation angle by the method used above showed that in this case $\Theta_{opt} = 0.7$, i.e., a factor of three less than the theoretical value. This shows that with autodyne detection, the signal maximum is caused not by a maximum in the value of $\overline{M}_y$ in the detector, but by a change in the sensitivity of the circuit.

Relation of the Signal Amplitude to the Strength and Nonuniformity of the Field in the Detector. According to expression (5.3), the signal amplitude is proportional to the strength of the field H_0 in the detector provided that $H_0 \ll H_p$ or with $H_0 \approx H_p$, if $\gamma^2 H_1^2 T_{1n} T_{2n} \gg 1$ and $v_a/aT_1 \ll 1$. Both of these requirements are normally fulfilled, i.e., theoretically, with a constant sensitivity of the detector circuit, A is proportional to H_a.

Let us examine the relation of the signal amplitude to the nonuniformity of the field H_a. It was established experimentally that the signal has a large amplitude if the gradient of the field H_0 is at right angles to the flow of the liquid. In this case the effect produced by the nonuniformity of the field is a decrease in the effective transverse relaxation time of the liquid, produced by dephasing of the precessing magnetic moments of the nuclei passing through the detector in different parts of the cross section.

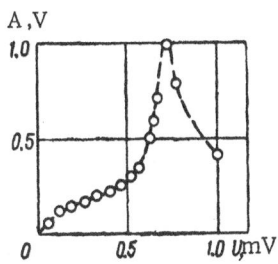

Fig. 7.3. Experimental relation of the nuclear resonance signal amplitude to the strength of the oscillating field with an autodyne detector used.

With a stationary working material, the presence of a field nonuniformity ΔH in the volume of the detector is known to reduce the transverse relaxation time to a value $T_2^* = 2/\gamma \Delta H$. As the dephasing of the magnetic moments of nuclei precessing in a nonuniform field is not distorted by movement of the nuclei in a direction at right angles to the field gradient, in a flow detector with nonuniformity of the field at right angles to the stream, the relaxation time $T_{2n} \approx 2/\gamma \Delta H_\perp$ (this is valid when there is no transverse mixing of the liquid in the stream, as is confirmed by the results of the experiment described in Section 2.2).

TABLE 1.3

$A \mid A_M$	10	15	10	10	15	15	10
H_0, oe	4	10	30	60	70	120	180
grad H_0, $\frac{oe}{cm}$	0.1	0.5	5	17	8	15	35
$\frac{grad\ H_0}{H_0}$, %	2.5	5	15	28	14	12	20

Thus, an expression for the signal amplitude in a nonuniform field may be obtained by substituting in expression (5.3) $T_{2n} = 2/\gamma \Delta H_\perp$. An investigation of the expression thus obtained, which is given in appendix 3, showed that the signal amplitude equaled 0.7 A_M on condition that $v_a \gamma \Delta H_\perp / 4q \ll 1$ and $(v_a/q)\, \gamma H_1 \approx 2.33$ (these conditions correspond to the case of low nonuniformity of the field).

With an increase in the nonuniformity of the field, the signal amplitude when $v_a \gamma \Delta H_\perp / 4q > 1$ is given with an error of less than 10% by

$$A = \frac{A_M}{(v_a \gamma \Delta H_\perp / 4q) + 1}, \tag{13.3}$$

i.e., in a nonuniform field the signal amplitude is inversely proportional to the gradient.

The relation (13.3) was checked experimentally on the apparatus illustrated in Fig. 2.3 with a detector 0.03 cc in volume (see Fig. 3a.3). The diameter of the detector d = 0.3 cm and the liquid flow rate q = 47 cc/sec.

Table 1.3 gives the signal-to-noise ratios with several field strengths and gradients. The field gradient was determined from the change in the field strength when the detector was shifted 1 cm in the direction of the gradient. It is difficult to assess the relation of A to H_0 and grad H_0 directly from this table as both values are varying at the same time, but the table does show that an adequate signal is obtained with a field strength of 4 oe and above and also in a field with a relative nonuniformity of up to 28% per cm.

For checking expression (13.3), Table 2.3 was compiled from the data in Table 1.3.

In the first line we give in relative units the signal amplitudes A for a field H_0 = 30 oe, obtained from the third line of Table 1.3 on the assumption that the noise amplitude is independent of the field strength H_0, while the signal amplitude is proportional to it. In the third line of Table 2.3 we give the product of the signal amplitude A and (a + 1), where a = $(v_a/4q)\,\gamma \cdot d$ grad $H_0 \approx 1.28$ grad H_0. A graph of the relation of A(solid line) and A(a + 1) (broken line) to the heterogeneity of the field in the detector is shown in Fig. 8.3. With grad

TABLE 2.3

A	75	45	10	6.4	3.75	1.67
grad H_0	0.1	0.5	5	8	15	35
$A(a+1)$	84	74	74	72	76	76.5

$H_0 > 1$ oe/cm, the broken line is parallel to the abscissa axis, i.e., with grad $H > 1$ oe/cm or $a > 1.3$, the signal amplitude $A \sim (a + 1)^{-1}$, which corresponds to expression (13.3).

The fact that with high nonuniformity of the field in the detector ΔH the signal amplitude is inversely proportional to ΔH may be explained from two points of view. (a) As already stated in the derivation of the theoretical relation in appendix 3, with nonuniformity ΔH of the rotating field of high strength H_1, not all the nuclei in the detector at a given moment participate in the resonance effect, but only the fraction of the nuclei lying in the part of it where the field strength differs from the resonance strength by a value of the order of the line width, which, as will be shown below, approximately equals $2H_1$. In other words, into the expression for the signal amplitude it is necessary to introduce a factor inversely proportional to ΔH, which characterizes the effective filling factor of the detector with resonating nuclei. (b) Another explanation may be put forward on more general bases: with high nonuniformity of the field, the width of the nuclear resonance line is proportional to ΔH and therefore its amplitude is inversely proportional to ΔH as the total intensity of the line, i.e., its area, is determined by the number of nuclei in the detector.

As a result of the investigation of the nuclear resonance signal amplitude, the following conclusions can be drawn:

1) The relation of the signal amplitude to the parameters of the detector is described satisfactorily by expression (5.3);

2) The signal amplitude increases with an increase in the detector volume and a good signal is obtained with a detector volume of 0.03 cc;

3) Provided that there is complete preliminary polarization $\gamma^2 H_1^2 T_{1n} T_{2n} \gg 1$; $v_a/qT_{2n} > 4$ and $(H_0/H_n) \cdot (v_a/qT_1) \ll 1$, the signal amplitude is proportional to the liquid flow rate and is independent of T_1;

4) With an increase in the strength of the oscillating field H_1, the signal amplitude increases to some optimal strength $H_1 = 2.36q/\gamma v_a$, at which the angle of nutation of the magnetization of the nuclei passing through the detectors equals $3\pi/4$. If the sensitivity of the detector circuit changes with an increase in H_1, this produces a corresponding shift in the optimal value of H_1.

5) With a change in the nonuniformity of the field in the volume of the detector ΔH, the signal amplitude is inversely proportional to ΔH when $\Delta H \gg 4q/\gamma \Delta H_0$ and independent of ΔH when $\Delta H \ll 4q/\gamma v_a$. A satisfactory signal is observed with a relative heterogeneity up to 20-30% per centimeter;

6) The signal amplitude is proportional to the strength of the field in the detector. With a detector volume of 0.03 cc, a satisfactory signal is observed with a field strength of 4 oe and above.

At this point it should be emphasized that if the topography (relative nonuniformity) of the field does not change with an increase in its strength, with a sufficiently high nonuniformity the signal amplitude is independent of the field strength as the effect of an increase in the strength is balanced by the effect of the increase in the absolute nonuniformity. Therefore, in practice, with a flow detector with the same polarizer approximately the same signal-to-noise ratio is obtained in fields from 10 to 5000 oe.

Width of Absorption Signal. The width of an absorption signal is determined by several different factors. The width of the nuclear resonance line is determined by the total probability of reorientation of the nuclei in the detector. The signal width increases with an increase in the percentage and frequency of modulation of the external magnetic field and with an increase in the nonuniformity of the external field in the volume of the detector. With the use of a radiofrequency circuit with a high Q-factor there is also radiation broadening of the

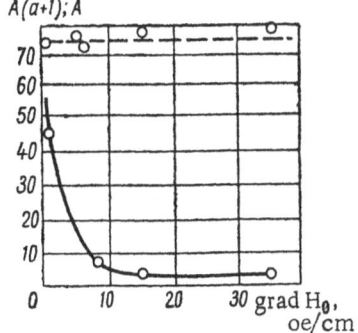

Fig. 8.3. Relation of the signal amplitude to the nonuniformity of the field in the detector: —— relation of A to grad H_0; ---- relation of $A(a + 1)$ to grad H_0.

nuclear resonance line. In addition to the normal reasons for signal broadening listed above, in a flow detector there is also so-called instrument broadening and broadening due to the finite time the nuclei remain in the volume of the detector. These are examined in Ch. 4. Let us estimate the width of the nuclear resonance line in a flow detector associated with the action of the radiofrequency field.

The absorption signal amplitude is proportional to the mean component M_y of the nuclear magnetization through the volume of the detector. This is maximal when the frequency of the oscillating field in the detector equals the precession frequency of the nuclei. If there is a slight shift $\Delta\omega$ between these frequencies, $\overline{M}_y$ decreases with an increase in $\Delta\omega$. The frequency shift at which $\overline{M}_y$ equals half of the maximum value is the half-width of the nuclear resonance line at the half-height. To determine this value it is necessary to find the relation of $\overline{M}_y$ to the frequency shift. The change in the projection M_y of the nuclear magnetization in a uniform external field under the influence of the oscillating field with a frequency shift $\Delta\omega$ is described by the expression

$$M_y = -\frac{Mp}{\sqrt{\frac{\Delta\omega^2}{\gamma^2 H_1^2}+1}}\, e^{-\frac{t}{T}} \sin\sqrt{\gamma^2 H_1^2 + \Delta\omega^2}\, t\,, \tag{14.3}$$

where H_1 is half the amplitude of the oscillating field and T is the relaxation time of the liquid (in a uniform external field $T_{1n} \approx T_{2n}$). This expression is obtained in appendix 4 by solving a system of equations describing the change in the magnetization of the nuclei under the influence of a rotating magnetic field in a rotating system of coordinates with $\Delta\omega \neq 0$ and $T_{1n} = T_{2n} = T$.

The relation (14.3) describes the change in M_y along the stream if the coordinate x, measured from the beginning of the detector along the stream, is related to time by $x = tW$, where W is the flow velocity of the liquid. To average M_y through the volume of the detector it is necessary to integrate expression (14.3) with respect to t within the limits of 0 and l_a/W. Then

$$\overline{M}_y = -\frac{Mp}{\gamma H_1 \frac{l_a}{W}\left(1+\frac{1}{T^2\gamma^2 H_1^2}+\frac{\Delta\omega^2}{\gamma^2 H_1^2}\right)}\left[1-e^{-\frac{l_a}{WT}}\left(\cos\frac{l_a}{W}\sqrt{\gamma^2 H_1^2+\Delta\omega^2}+\frac{\sin\frac{l_a}{W}\sqrt{\gamma^2 H_1^2+\Delta\omega^2}}{T\sqrt{\gamma^2 H_1^2+\Delta\omega^2}}\right)\right]. \tag{15.3}$$

In addition to $\Delta\omega$, this expression includes H_1, which may be set arbitrarily. It is most advantageous to select H_1 by using condition (8.3), guaranteeing the maximum value of $\overline{M}$ with accurate resonance tuning, i.e., $H_1 = 3\pi W/4\gamma l_a$. As normally $l_a/W \ll T$, the inequality $\gamma H_1 T \gg 1$ holds. In this case the broadening due to spin-spin and spin-lattice interaction is negligibly small and expression (15.3) has the form

$$\overline{M}_y = -\frac{Mp}{\frac{3\pi}{4}\left(1+\frac{\Delta\omega^2}{\gamma^2 H_1^2}\right)}\left(1-\cos\frac{3\pi}{4}\sqrt{1+\frac{\Delta\omega^2}{\gamma^2 H_1^2}}\right). \tag{16.3}$$

This relation is illustrated in Fig. 9.3. The figure shows that $\overline{M}_y/M_p$ falls by a factor of 2 when the frequency difference $\Delta\omega = 1.1\gamma H_1$, hence the nuclear resonance line width at the half-height with the optimal amplitude of the oscillating field

$$\delta\omega_a = 2.2\gamma H_1. \tag{17.3}$$

This line width which is associated with the action of the radiofrequency field may be observed only when a bridge circuit is used to detect the signal. Autodyne circuits normally have maximal sensitivity at a low oscillation amplitude and therefore, in work with them the optimal amplitude of the oscillation field is considerably less than the value determined with expression (8.3) and the line broadening produced by it is small.

Let us estimate the broadening due to the finite residence time of the nuclei in the detector. During the passage of the nuclei through the detector there actually acts on them a pulse of the resonance oscillating field, the form of which depends on the topography of the field of the detector coil and the duration of the pulse is determined by the value v_a/q, where v_a is the volume of liquid in the coil and q is the liquid flow rate. For example, if the oscillating field is identical throughout the volume v_a and it falls sharply beyond its limits, then the pulse is rectangular. In this case, instead of the fundamental frequency of the oscillating field, there acts on the nuclei a continuous spectrum of frequencies with a half-width at the half-height equal to q/v_a Hz. This produces a corresponding boradening of the nuclear resonance line proportional to the liquid flow rate.

As will be shown in Ch. 4, the so-called instrument broadening, which is present in detectors of some designs, is also connected with the movement of the nuclei and has a value of the order of q/v_a Hz, i.e., the signal broadening produced by the movement of the nuclei through the detector is determined by the expression

$$\delta f = k \frac{q}{v_a},$$ (18.3)

where k is a numerical coefficient, depending on the geometry of the detector coil.

The factors in the broadening of a nuclear resonance line examined play a part only with a sufficiently uniform external field, i.e., if its nonuniformity within the detector $\Delta H \ll q/\gamma v_a$. Otherwise the width of the nuclear resonance line is determined by the nonuniformity of the field in the detector.

For an experimental investigation of the width of a nuclear resonance line we used the apparatus illustrated in Fig. 2.3. It included an iron-clad polarizing magnet with a field strength of 10,000 oe and a chamber volume of 400 cc, and a connecting tube with a diameter of 0.4 cm and a length of 100 cm. The detector is shown in Fig. 3a.3. The working substance was water from a tap. Resonance was observed with a field strength of about 10 oe with an autodyne detector circuit [59].

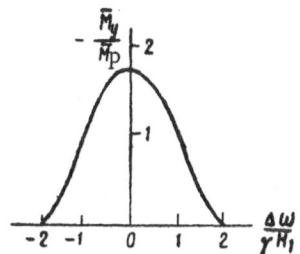

Fig. 9.3. Theoretical relation of the absorption signal amplitude to the frequency difference.

The width of the nuclear absorption signal was determined with a set percentage of low-frequency modulation of the external field. The maximum of the second harmonic of the signal (with a frequency equal to twice the frequency of the modulating field) was obtained exactly at resonance when the percentage modulation equaled the width of the absorption line, while the first harmonic (with a frequency equal to the modulation frequency of the field) had two maxima in antiphase, at a space equal to the percentage modulation greater than or equal to the width of the absorption line.

Thus, to measure the width of a nuclear resonance line it is necessary to set the minimum percentage modulation of the external field at which the amplitude of the first harmonic of the signal is still maximal, to select and measure the frequency of the oscillating field at which the maximum signal is observed, then to shift the frequency until the appearance of the maximum signal in antiphase and again measure it. The difference in the frequencies in hertz equals the width of the line δf.

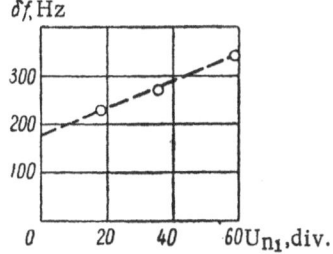

Fig. 10.3. Experimental relation of the width of the nuclear absorption signal to the strength of the oscillating field.

The relation of δf measured in this way in a uniform field to the strength of the oscillating field is shown in Fig. 10.3. The maximum potential corresponds to the optimal oscillation frequency of the autodyne circuit used. The figure shows that with an autodyne detector, as was expected, the maximum broadening produced by the oscillating field was no more than 30% of the line width at U_{n_1}, which equaled 180 Hz.

The nonuniformity of the field in the volume of the detector was of the order of 0.001 oe, i.e., the contribution to the line width product by it did not exceed 4-5Hz.

The absence of appreciable radiation broadening was confirmed by the form of the signal observed. Consequently, the width of the signal at $U_{n_1} = 0$ was caused by the instrument effect and the finite time for the nuclei to pass through the detector.

49

The detector volume v_a = 0.028 cc and the flow rate q = 40 cc/sec. Knowing these values it was possible to determine the proportionality k between the total width of the signal δf caused by the movement of the nuclei through the detector and q/v_a. In the case of a cylindrical flow detector we found that k = 1/8, i.e., δf =(1/8) q/v_a Hz.

The following experiment was designed to determine the relation of the line width to the nonuniformity of the field in the volume of the detector. In a field of known nonuniformity the first harmonic of the signal was detected and the field strength set exactly at resonance (zero signal amplitude). The detector was then shifted in the direction of the maximum gradient until there appeared a signal of maximum amplitude in phase with the modulation and then the detector was shifted in the opposite direction until there appeared the maximum signal in antiphase. The field had a constant gradient along the course over which the detector was moved and therefore the width of the nuclear absorption line $\delta\omega$ = γ grad $H_0 \Delta l$, where Δl is the distance between the positions of the detector in which the maxima of the signals in antiphase were observed. In all cases with the optimal percentage modulation this distance was approximately equal to half the linear dimension of the detector: $\Delta l \approx d_a/2$.

Thus, $\delta\omega$ = γ grad $H_0(d_a/2)$, i.e., the width of the nuclear resonance line in a nonuniform field equals half the nonuniformity of the field in the volume of the detector.

2.3. Nuclear Induction Signal

The two-coil method, proposed by Bloch, Hansen, and Packard [76, 77], is usually used to observe a nuclear induction signal. In this method the rotation of the magnetization and the observation of the signal induced in the circuit by the precessing nuclei are achieved with different coils, namely, a transmitting coil, which is connected to a radiofrequency oscillator, and a receiving coil, whose output is connected to a highly sensitive amplifier.

The most important problem in these detectors is decoupling of the receiving and transmitting coils so that the resonance oscillating magnetic field excited in the transmitting coil does not induce an appreciable emf in the receiving coil. If this is not done the emf produced by the precessing magnetization of the nuclei is difficult to measure on the pickup background. For this purpose the axes of the coils lie at an angle of 90°, the direction of the magnetic flux of the transmitting coil is controlled by various mechanical devices, and electronic compensation is also used [76, 77, 88-91]. The problem is considerably simplified in using the induction method for recording the nuclear resonance signal in a flow detector as in this case the transmitting and receiving coils may be separated in space and carefully screened from each other. Such a detector was first constructed by Sherman [4, 5]. A diagram of the detector is given in Fig. 11.3. It was made in the form of an aluminum chamber with compartments for the transmitting and receiving coils and their trimmers. The liquid flowed into the chamber through a polyethylene tube 1, with an internal diameter of 0.275 mm and an external diameter of 0.6 mm. The tube passed through the transmitting coil 2 and the isolating wall 3 into the compartment where the receiving coil 5 lay, passed through the latter, and led to a connecting pipe for draining off

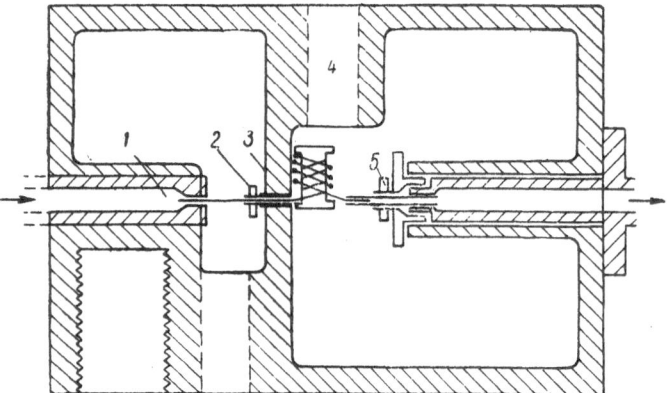

Fig. 11.3. Diagram of nuclear induction flow detector: 1) inlet tube; 2) transmitting coil; 3) isolating wall; 4) cable connector for receiving coil; 5) receiving coil.

the liquid. In the design described, provision was made for varying the length of the tube between the transmitting and receiving coils for determining its effect. The transmitting coil had 20 turns of copper wire, 0.055 mm in diameter, wound onto a form 1 mm in diameter. Its Q-factor was 26 at a frequency of 15 MHz. The receiving coil had 15 turns of copper wire, 0.082 mm in diameter, wound onto a form 1.4 mm in diameter. Its Q-factor was 33 at a frequency of 15 MHz. Both coils were matched with coaxial lines by means of special trimmers. The chamber was placed in the interpolar space of a magnet with a pole diameter of 30 cm, a gap of 4.4 cm, and a magnetic field strength of about 3500 oe. To obtain a nuclear resonance signal, an oscillating magnetic field with a frequency of 15 MHz was excited in the transmitting coil and the magnetic field strength slowly varied.

When the field strength passed through the resonance value, a nuclear induction signal appeared at the output of the detecting system. The most interesting result obtained in this experiment was the fact that the frequency of the emf induced in the receiving coil by the precessing magnetization of the nuclei did not equal the precession frequency of the nuclei, but equalled the frequency of the oscillating magnetic field in the transmitting coil. This fact, which is paradoxical at first glance, is readily explained by an elementary examination of the processes in the detector.

An oscillating magnetic field is known to be equivalent to the sum of two fields of half the amplitude rotating in opposite directions. The component rotating in the same direction as the precessing magnetic moments of the nuclei is active in nuclear magnetic resonance. If the frequency of the rotating field exactly equals the Larmor precession frequency of the nuclei, then there is rotation of the magnetization of the nuclei in a plane normal to the vector of the rotating field, i.e., the phase of the precessing magnetization differs by $\pi/2$ from the phase of the rotating field. In the more general case, when the frequencies of the rotating field and the nuclei differ by $\Delta\omega$, after deflection of the magnetization by the rotating field, its phase may differ from the phase of the rotating field by some value $\Delta\varphi$, which depends on $\Delta\omega$, H_1, and τ, where H_1 is the vector of the rotating field and τ the time of action of the field H_1 on the nuclei. If an alternating potential with a frequency ω_0 is applied to the transmitting coil, then the phase of the rotating field changes with time t according to the relation $\varphi = \omega_0 t$ and the phase of the magnetization of the nuclei in the liquid emerging from the transmitting coil at a moment of time t has the value $\varphi_1 = \omega_0 t + \Delta\varphi$. The value of $\Delta\varphi$ is independent of time.

In the subsequent flow of the liquid, the phase of the precessing magnetization changes with the Larmor precession frequency of the nuclei. If on the course of the nuclei between the transmitting and receiving coils the magnetic field changes accoding to the law H(x) (the coordinate x is measured from the end of the transmitting coil along the stream of the liquid), the phase of the magnetization in the liquid flowing into the receiving coil will have the value

$$\varphi_1 = \omega_0 t + \Delta\varphi + \int_0^l [\gamma H(x) - \omega_0] \frac{dx}{W}, \tag{19.3}$$

where l is the distance between the coils and W is the velocity of the liquid. The frequency of the emf induced by the nuclei in the receiving coil equals the frequency of rotation of magnetization of the nuclei. In the expression for φ_1, only the first term depends on time, i.e., $\omega = \omega_0$.

The amplitude of the induction emf of the nuclei is proportional to $\cos(\varphi_1 + \alpha)$, where α is some phase angle depending on the angle between the axes of the transmitting and receiving coils. The value of φ_1 varies along the length of the receiving coil and therefore, to find the expression for the amplitude of the induction signal it is necessary to integrate

$$A\varphi_{1l} = \frac{k}{\varphi_{1L}} \int_{\varphi_{1l}}^{\varphi_{1L}} \cos(\varphi_1 + \alpha) d\varphi_1 = \frac{\sin(\varphi_{1L} + \alpha) - \sin(\varphi_{1l} + \alpha)}{\varphi_{1l} - \varphi_{1L}}, \tag{20.3}$$

where φ_{1l} and φ_{1L} are the values of φ_1 at the beginning and the end of the receiving coil at the moment that the signal is observed and k is the proportionality coefficient. By using the expression (19.3) we have

$$\varphi_{1l} = \omega_0 t + \Delta\varphi + \int_0^l [\gamma H(x) - \omega_0] \frac{dx}{W} = \omega_0 t + \Delta\varphi + (\gamma\overline{H}_l - \omega_0)\frac{l}{W} , \qquad (21.3)$$

$$\varphi_{1L} = \omega_0 t + \Delta\varphi + \int_0^{l+L} [\gamma H(x) - \omega_0] \frac{dx}{W} = \omega_0 t + \Delta\varphi + (\gamma\overline{H}_l - \omega_0)\frac{l}{W} + (\gamma\overline{H}_L - \omega_0)\frac{L}{W} , \quad (22.3)$$

where L is the length of the receiving coil, $\overline{H}_1$ the mean magnetic field strength at a distance l, and $\overline{H}_L$ is the mean magnetic field strength in the receiving coil.

By substituting these values in expression (20.3) and carrying out trigonometric rearrangements we obtain the relation of the nuclear induction emf to the various parameters of the apparatus

$$A = 2k \cos\left(\omega_0 t + \Delta\varphi + \gamma\overline{H}_l \frac{l}{W} + \gamma\overline{H}_L \frac{L}{2W} - \omega_0 \frac{l + \frac{L}{2}}{W}\right) \frac{\sin(\gamma\overline{H}_L - \omega_0)\frac{L}{2W}}{(\gamma\overline{H}_L - \omega_0)\frac{L}{2W}} . \qquad (23.3)$$

In the experiment described by Sherman, a magnet with a highly uniform field was used, i.e., $\overline{H}_L = \overline{H}_l = H$, moreover, the frequency ω_0 was close to the precession frequency of the nuclei in the transmitting coil $\Delta\varphi = \pi/2$. Then expression (23.3) is simplified

$$A = 2k \sin\left[\omega_0 t + (\gamma H - \omega_0)\frac{l + \frac{L}{2}}{W}\right] \frac{\sin(\gamma H - \omega_0)\frac{L}{2W}}{(\gamma H - \omega_0)\frac{L}{2W}} . \qquad (24.3)$$

In observing a nuclear resonance signal it is usual to separate the signal arising in phase with the potential in the transmitting coil, which is called the dispersion signal, and the signal arising in quadrature with it, which is called the absorption signal. Sherman observed the absorption signal, whose amplitude from expression (24.3) is proportional to

$$\cos\left[(\gamma H - \omega_0)\frac{l + \frac{L}{2}}{W}\right] \frac{\sin(\gamma H - \omega_0)\frac{L}{2W}}{(\gamma H - \omega_0)\frac{L}{2W}} . \qquad (25.3)$$

The experimental relation of the nuclear induction signal to the strength of the magnetic field H is given in Fig. 12.3. One division along the abscissa axis corresponds to a change in the field of 0.0425 oe or in units of proton resonance frequency 180 Hz. The signal consists of a series of oscillations with a period equal to 30 Hz. In one division of the graph there are 6 signal periods. This type of signal is caused by the first factor in expression (25.3). From this factor, the oscillation period must equal $1/(l/W + L/2W)$, which equals 33 Hz under the experimental conditions. The discrepancy is within the limits of experimental error. The envelope signal is determined by the second cofactor of expression (25.3), whence the width of the maximum of the envelope signal between the points at which the signal equals zero must equal 2W/L Hz. Under the experimental conditions, 2W/L is 360 Hz. In Fig. 12.3 the width of the maximum of the envelope signal approximately equals two divisions which exactly corresponds to the theoretical value. In practice, the fall of the envelope signal occurs more rapidly than follows from expression (25.3). This may be explained by the nonuniformity of the liquid velocity curve across the

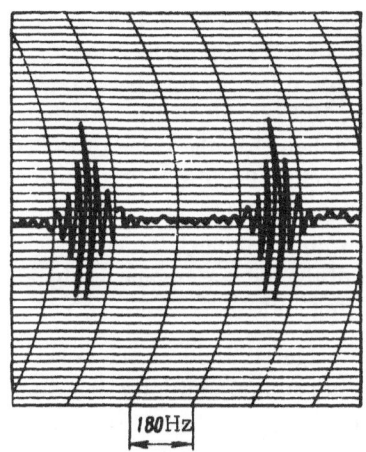

Fig. 12.3. Experimental form of the nuclear induction signal in a flow detector with l/W - 0.03 sec.

section of the tube. A more rapid fall of the envelope signal may be obtained by using modulation of liquid flow velocity. The nuclear resonance signal without modulation of the flow with l/W = 0.1 sec is given in Fig. 13a.3; the same signal with modulation of the flow is given in Fig. 13b.3. The modulation frequency equaled 1.5 Hz and the ratio of the maximum and minimum velocities was two. The ratio of the amplitude of the central peak to the amplitude of the first side peaks in the case of Fig. 13a.3 is 1.35 and in the case of Fig. 13b.3, 2.6, i.e., with modulation of the flow the central peak stands out much more sharply. The narrowest nuclear resonance line obtained on the apparatus described is shown in Fig. 14.3. The width of the central peak is 1.4 Hz, i.e., 10^{-7} in relative units. This result was obtained with the length of the tube connecting the transmitting and receiving coils l = 10 cm and l/W = 0.244 sec. A further increase in the time for the nuclei to pass between the coils produces a con-- siderable fall in the signal-to-noise ratio. Graphs of two envelopes of nuclear resonance signals with l/W = 0.1 sec are given in Fig. 15.3. The symmetrical curve 1 was obtained when the detector was in the most uniform part of the field of the magnet. The asymmetric curve 2 was obtained when the field had a gradient of 0.7 oe/cm. The connecting tube l was at a distance of 0.5 cm from the receiving coil, i.e., in a field which differed from that in the receiving coil by 0.35 oe (150 Hz).

A nuclear induction flow detector was used by F. I. Skripov to measure the earth's magnetic field [13]. In this apparatus the liquid was polarized by passing through a space inside a system of Helmholtz coils and then it passed through a transmitting coil, where under the influence of a resonance oscillating magnetic field the magnetization of the nuclei was deflected through an angle of the order of 90° from the direction of the measuring field, and into a receiving coil, where the precessing magnetization of the nuclei induced an emf of free nuclear induction.

The calculated amplitude-frequency and frequency-phase characteristics of such an apparatus [61] have the form

$$\frac{A(\delta\omega)}{B(0)} = \sqrt{\frac{1+\dfrac{2e^{-\frac{\tau_r}{T_2}}}{(1-e^{-\frac{\tau_r}{T_2}})^2}(1-\cos\tau_r\,\delta\omega)}{(1+T_2\delta\omega)^2}}, \qquad (26.3)$$

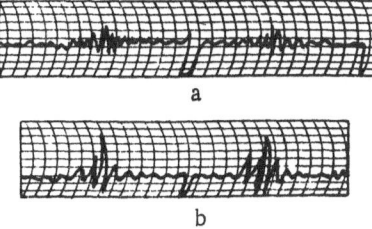

Fig. 13.3. Experimental form of nuclear induction signal in a flow detector with l/W = 0.1 sec: a) without modulation of flow; b) with modulation of flow.

$$\Phi = \frac{1-\cos\gamma H_1\tau_\phi}{\gamma H_1\sin\gamma H_1\tau_\phi} + \delta\omega\tau_c + \tan^{-1}T_2\delta\omega -$$

$$- \tan^{-1}\frac{e^{-\frac{\tau_r}{T_2}}\sin\delta\omega\tau_r}{1-e^{-\frac{\tau_r}{T_2}}\cos\delta\omega\tau_r}, \qquad (27.3)$$

where A is the signal amplitude, Φ its phase, $\delta\omega = \omega_0-\omega$ is the difference in the frequency ω of the oscillating field in the transmitting coil and the resonance frequency, γ is the gyromagnetic ratio, H_1 is the half-amplitude of the radiofrequency field in

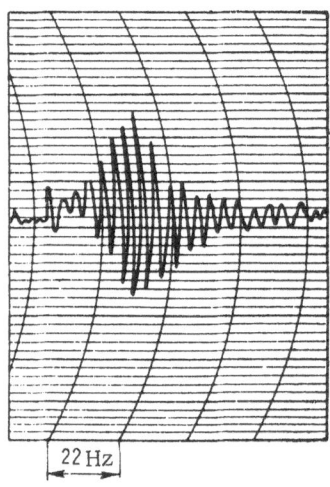

Fig. 14.3. Experimental form of nuclear induction signal in a flow detector with $l/W = 0.244$ sec.

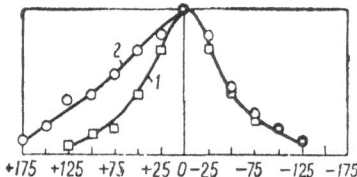

+175 +125 +75 +25 0 -25 -75 -125 -175

Fig. 15.3. Experimental form of nuclear induction signal in a flow detector with $l/W = 0.1$ sec: 1) uniform field; 2) field with grad H = 0.7 oe/cm.

the transmitting coil, T_2 is the transverse relaxation time, and τ_Φ, τ_c, and τ_r are the times for flow through the field H_1, the connecting tube, and the receiving coil, respectively. These relations are given graphically for various values of τ_r/T_2 in Figs. 16.3 and 17.3. The relation of the half-width of the signal at the half-height to the value of τ_r/T_2 obtained from the graph in Fig. 16.3 is given in Fig. 18.3. These values give an idea of the character of the broadening of the nuclear resonance line with an increase in the liquid flow velocity through the detector and the related phase shifts of the potential of the nuclear resonance signal relative to the potential in the transmitting coil.

For accurate measurements of the magnetic field, to the transmitting coil is fed a potential not from an external oscillator, but from the output of the amplifying apparatus connected to the receiving coil. Such a system is capable of self-excitation with a frequency close to the frequency of nuclear precession in the transmitting and receiving coils. So that the polarizing field did not distort the results of the measurements, the Helmholtz coils were designed for compensation of the stray field in the region of the receiving coil to a value of the order of 10^{-6} oe. F. I. Skripov did not report [61] the possible accuracy of the measurements of the earth's magnetic field.

3.3. Nuclear Resonance Oscillators. In previous sections we examined two different methods of observing a nuclear resonance signal. For observing an absorption signal, a substance containing nuclei with magnetization oriented in the direction of the external field must be placed in the coil of radiofrequency circuit and a magnetic field excited in it with a frequency close to the nuclear precession frequency. As a result of the exchange of energy of the nuclei with the oscillating magnetic field, energy is absorbed from the circuit and this produces a fall in its Q-factor, which is recorded.

For observing an induction signal, a substance with the magnetization of the nuclei oriented at right angles to the external field must be placed in the coil of the circuit. This state

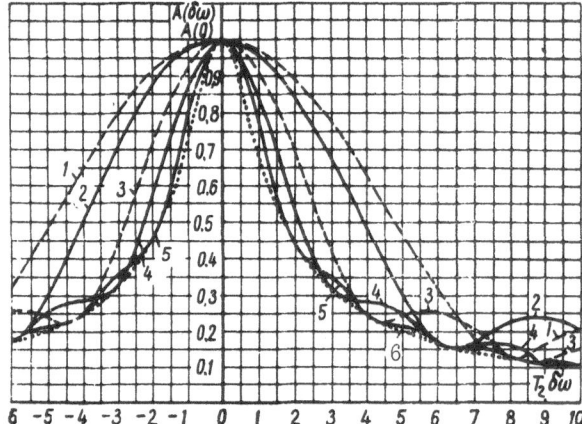

Fig. 16.3. Theoretical relation of the amplitude of the nuclear induction signal in a flow detector to the frequency of the oscillating field in the transmitting coil: 1) $\tau_r/T_2 = 0.8$; 2) $\tau_r/T_2 = 1$; 3) $\tau_r/T_2 = 1.5$; 4) $\tau_r/T_2 = 2$; 5) $\tau_r/T_2 = 3$; 6) $\tau_r/T_2 \rightarrow \infty$.

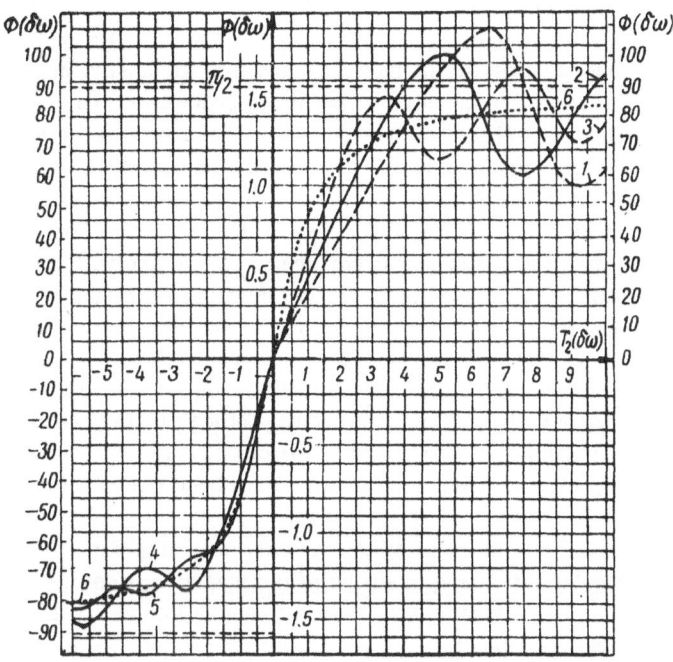

Fig. 17.3 Theoretical relation of the phase of the nuclear induction signal in a flow detector to the frequency of the oscillating field in the transmitting coil: 1) $\tau_r/T_2 = 0.8$; 2) $\tau_r/T_2 = 1$; 3) $\tau_r/T_2 = 1.5$; 4) $\tau_r/T_2 = 2$; 5) $\tau_r/T_2 = 3$; 6) $\tau_r/T_2 \to \infty$.

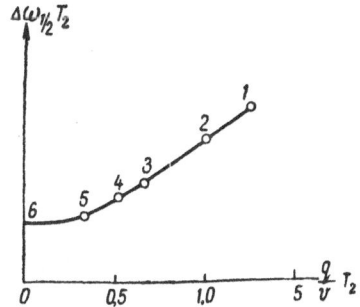

Fig. 18.3. Theoretical relation of the width of the nuclear induction signal in a flow detector to the liquid flow velocity through the detector.

of the substance is unstable and therefore it tends to change to a stable state with the transverse component of the magnetization equal to zero by transferring the excess energy to the circuit. As a result of this, a nuclear induction emf is induced in the circuit. As a result of this, a nuclear induction emf is induced in the circuit and the magnetization of the nuclei falls with a relaxation time

$$\tau_r = \frac{1}{2\pi\eta\gamma QM},$$

where η is the filling factor of the coil of the circuit with nuclei, Q is the quality factor of the circuit, and γ is the gyromagnetic ratio of the nuclei. This addition damping of the nuclear magnetization as a result of the radiation of energy into the radiofrequency coil by the nuclei was first studied by Blombergen and Pound [92] and was called "radiation damping." Blombergen and Pound denied the possibility of radiation damping with the magnetization of the nuclei oriented opposite to the external field.

However, a calculation carried out by K. V. Vladimirskii showed that radiation damping may also be observed with negative magnetization of the nuclei if the condition $T_2 > \tau_r$ holds (T_2 is the transverse relaxation time of the nuclei.

Let us examine the excitation of a "maser." The presence in the coil of a radiofrequency circuit of nuclei with their magnetization oriented along the external field does not produce oscillations in it. However, if the circuit is tuned to a frequency close to the nuclear precession frequency, in the spectrum of its noise a large fraction consists of noise with the frequency of the nuclear precession. Under the influence of this noise, the magnetization will be deflected slightly from the direction of the external field. Thereupon there appear transverse components of the nuclear magnetization which induce an emf in the circuit. If the magnetization of the nuclei is oriented in the direction of the external field, this induced emf arises in antiphase with the noise fluctuations which were responsible for the appearance of the transverse components. If the magnetization is oriented opposite to the external field, this emf is produced in phase and with a sufficiently high Q-factor of the circuit, it produces a still greater deviation of the magnetization, which leads to an increase in the transverse components and, consequently, to a further increase in the amplitude of the oscillations in the circuit.

Thus, if a circuit with a sufficiently high Q-factor is placed in a uniform magnetic field and a substance with negative polarization of the nuclei is passed continuously through it, then oscillations with a frequency close to the nuclear precession frequency are produced in the circuit. In analogy with Towns molecular oscillator with a beam of ammonia molecules [94], such systems have been called "nuclear resonance masers."

At the present time there are two methods of continuously obtaining nuclei with negative polarization (see Ch. 2), namely, dynamic polarization and inversion of the magnetization of the nuclei in a polarized flowing liquid. Using the first method, Abraham, Cambrisson, and Solomon [57] built a "maser" in which the working substance was an aqueous solution of potassium nitroso-disulfonate (Fremy's salt). A "maser" was built by the French scientist Benoit [81] on the basis of a nuclear resonance flow detector. Let us examine in more detail the characteristics of the operation of a nuclear resonance "maser."

The effect of the polarized nuclei in the circuit is equivalent to the action of a complex dynamic magnetic susceptibility $X = X' - iX''$, which changes its inductance by the value $\Delta L = 4\pi\eta LX$, where L is the inductance in the absence of the nuclei.

The impedance of such a circuit

$$Z = r - \frac{i}{\omega C} + i\,\omega L\,[1 + 4\pi\eta\,(X' - iX'')], \qquad (28.3)$$

where r and C are the effective resistance and capacitance of the circuit. The active component of the resistance

$$R = r + 4\pi\eta\omega LX'', \qquad (29.3)$$

and the reactive component of the resistance

$$X = \omega L - \frac{1}{\omega C} + 4\pi\eta\omega LX'. \qquad (30.3)$$

When the condition for amplitude balance holds

$$R = 0 \qquad (31.3)$$

and the condition for the phase balance

$$X = 0 \qquad (32.3)$$

self-excitation of the system occurs.

By substituting formula (29.3) in condition (31.3) and formula (30.3) in (32.3) we obtain the conditions for self-excitation in a more expanded form

$$r + 4\pi\eta\omega L X'' = 0, \tag{33.3}$$

$$\omega L - \frac{1}{\omega C} + 4\pi\eta\omega L X' = 0. \tag{34.3}$$

The expressions for X' and X" for the condition of low saturation are known from Bloch's theory. For the case of a detector with extraneous polarization, in these expressions it is necessary to replace $X_0\omega_0$ by γM_p, where X_0 is the static nuclear magnetic susceptibility, ω_0 is the precession frequency of the nuclei in the external field, and M_p is the vector of the polarization of the nuclei entering the detector. After this replacement, these expressions have the form:

$$X'' = \frac{\gamma M_p T_2}{2\left[1 + (\omega - \omega_0)^2 T_2^2\right]}, \tag{35.3}$$

$$X' = \frac{\gamma M_p T_2^2 (\omega - \omega_0)}{2\left[1 + (\omega - \omega_0)^2 T_2^2\right]}. \tag{36.3}$$

By substituting X" from (35.3) in formula (33.3) we obtain

$$\frac{2\pi\eta Q\gamma M T_2}{1 + (\omega - \omega_0)^2 T_2^2} = -1. \tag{37.3}$$

Expression (37.3) is the condition for operation of the "maser" analogous to that obtained by K. V. Vladimirskii [93], but with allowance for the pulling effect, as a result of which the oscillation frequency of the "maser" ω differs slightly from the nuclear precession frequency ω_0. Let us determine the magnitude of this effect.

By substituting X' from expression (36.3) in formula (34.3) we obtain

$$\omega = \frac{\dfrac{1}{\sqrt{LC}}}{\sqrt{1 - \dfrac{2\pi\eta\gamma M_p T_2^2 (\omega - \omega_0)}{1 + (\omega - \omega_0)^2 T_2^2}}}. \tag{38.3}$$

As the frequency of the oscillations in the circuit excited by the nuclei differs little from its natural resonance frequency $\omega_p = 1/\sqrt{LC}$, the value under the root sign is close to 1 and expression (38.3) may be simplified

$$\omega = \omega_r \left[1 + \frac{\pi\eta\gamma M_p T_2^2 (\omega - \omega_0)}{1 + (\omega - \omega_0)^2 T_2^2} \right]. \tag{39.3}$$

By substituting in expression (39.3) the limiting value of M_p from formula (37.3) we obtain

$$\omega = \omega_r \left[1 - \frac{T_2}{2Q} (\omega - \omega_0) \right]. \tag{40.3}$$

From expression (40.3) we may obtain the ratio of the frequency shifts of the oscillations of the system nuclei-resonance circuit from the resonance frequency of the circuit ($\omega - \omega_r$) and from the Larmor precession frequency of the nuclei ($\omega - \omega_0$)

$$\frac{\omega - \omega_r}{\omega - \omega_0} = -\frac{\omega_r T_2}{2Q} = -K. \tag{41.3}$$

The value ω_r/Q is the width of the resonance curve of the circuit at the half-height, while $2/T_2$ is the width of the nuclear absorption line, i.e., the coefficient K equals the ratio of the widths of the resonance curve of the circuit and the nuclear resonance line.

From formula (41.3) it is possible to find the relation of the frequency shifts $(\omega - \omega_r)$ and $(\omega - \omega_0)$:

$$\omega - \omega_r = -\frac{K}{K+1}(\omega_r - \omega_0), \tag{42.3}$$

$$\omega - \omega_0 = \frac{1}{K+1}(\omega_r - \omega_0). \tag{43.3}$$

The expressions obtained completely characterize the pulling effect. From them it follows that the oscillation frequency of the "maser" is intermediate between the natural frequency of the circuit and the nuclear precession frequency. If the width of the nuclear resonance line is considerably greater than the width of the resonance curve of the circuit, i.e., $K \gg 1$, then the oscillation frequency of the "maser" is close to the natural frequency of the circuit. If the nuclear resonance line is narrower than the resonance curve of the circuit, i.e., $K \ll 1$, the oscillation frequency of the "maser" is close to the nuclear precession frequency. Thus, in the measurement of a magnetic field with a "maser," it is necessary to guarantee that the nuclear resonance line is narrow and to use a circuit with the lowest possible Q-factor.

Let us examine the range of operation of a "maser." From expression (37.3) it is possible to determine the maximum difference between the oscillation frequency of the "maser" and the nuclear precession frequency with given parameters of the apparatus:

$$\omega - \omega_0 = \frac{1}{T_2}\sqrt{2\pi\eta\gamma Q|M_p|T_2 - 1}. \tag{44.3}$$

It is more interesting to know the maximum difference between the natural frequency of the circuit and the nuclear precession frequency at which operation of the "maser" is possible. For this it is necessary to use relation (43.3) and replace $(\omega - \omega_0)$ by $(\omega_r - \omega_0)$ in expression (44.3)

$$\omega_r - \omega_0 = \left(\frac{\omega}{2Q} + \frac{1}{T_2}\right)\sqrt{2\pi\eta\gamma Q|M_p|T_2 - 1}. \tag{45.3}$$

The expression in brackets is the sum of the half-widths of the resonance curve of the circuit and the nuclear resonance line. This relation shows that if we have the condition $2\pi\eta\gamma Q|M_p|T_2 = 1$, i.e., $T_2 = \tau_r$, then self-excitation of the "maser" can occur only if the natural frequency of the circuit and the nuclear precession frequency are equal. In the case of $2\pi\gamma\eta Q|M_p|T_2 > 1$ or $T_2 > \tau_r$, the range of operation of the "maser" is increased and when $T_2 \approx 5\tau_r$ it equals the sum of the widths of the resonance curve of the circuit and the nuclear absorption line.

Let us determine the amplitude of the oscillations in the circuit A with exact tuning of the circuit to the nuclear precession frequency. If the oscillation frequency and the inductance equal ω and L, then

$$A = \omega L I,$$

where I is the current in the circuit, $I = E/r$, here E is the emf induced in the circuit by the precessing magnetic moments of the nuclei, while r is the active resistance of the coil.

Thus

$$A = \frac{\omega L}{r} E = QE. \tag{46.3}$$

The strength of the rotating magnetic field in the coil is proportional to the current:

$$H_1 = \frac{LI}{2NS},$$

where N and S are the number of turns and the cross section of the coil. By expression I in terms of A we obtain the relation of H_1 to the amplitude of the oscillations in the circuit

$$H_1 = \frac{A}{2N\omega S}. \tag{47.3}$$

The magnetic induction created by the nuclei in the coil

$$B = 4\pi M_\perp e^{i\omega t},$$

where $M_\perp e^{i\omega t}$ is the component of the magnetic moment of the nuclei in the detector directed at right angles to the external field.

The magnetic flux through the section of the coil equals

$$\Phi = 4\pi \eta N S M_\perp e^{i\omega t},$$

where η is the filling factor. The amplitude of the emf induced by the nuclei in the circuit

$$E = -\frac{d\Phi}{dt} = 4\pi \eta N S \omega M_\perp e^{i\omega t - \frac{\pi}{2}}. \tag{48.3}$$

As the circuit is tuned for resonance with the oscillation frequency, the current in the circuit and, consequently, the oscillating magnetic field is in phase with E. According to expression (48.3), the phase of the emf differs from the phase of the rotating magnetic moment by $\pi/2$ and therefore, in this expression, instead of $M_\perp$ it is possible to write M_y, where, as was stipulated in Section 1.3, M_y denotes the transverse component of the magnetization, rotating in quadrature with H_1. Having carried out this substitution and using relation (46.3), we obtain the expression for the amplitude of the oscillations in the circuit

$$A = 4\pi \eta S N Q \omega M_y. \tag{49.3}$$

It can be shown that this expression is identical to expression (23) for the amplitude of the absorption signal obtained in appendix 2. In the detection of the absorption signal, the strength of the field H_1, which determines the value of M_y in the detector, is arbitrarily set by the level of the oscillations of the external oscillator, while in the case of a "maser," the strength of the field H_1 is related to the potential of the oscillations excited in the circuit by the nuclei themselves. In both cases the relation of M_y to H_1 and the other parameters of the detector is determined by expression (4.3), i.e., the expression for the amplitude of the nuclear resonance signal (5.3) is valid for the amplitude of the oscillations in the "maser" circuit A if we substitute in it the value of the strength of the rotating magnetic field H_1, expressed in terms of the amplitude A in formula (47.3), when we will have the relation of A to the parameters of the "maser" in an implicit form. Let us do this for the case when the condition (6.3) for obtaining the maximum nuclear resonance signal holds. In this case the relation of the nuclear resonance signal amplitude to H_1 is determined by the expression (7.3):

$$A = 4\pi\eta NSQ\omega \, |M_p| \, \frac{1 - \cos \gamma H_1 \tau}{\gamma H_1 \tau} \, , \tag{50.3}$$

where $|M_p|$ is the magnetization of the nuclei in the liquid flowing into the detector coil, $\tau = v/q$ is the time for the nuclei to pass through the detector, v is the volume of the detector, and q the liquid flow rate. After substituting in this expression the value $H_1 = A/2N\omega S$ we obtain

$$A = 4\pi\eta NQS\omega \, |M_p| \, \frac{1 - \cos \dfrac{A\tau}{2NHS}}{\dfrac{A\tau}{2NHS}} \, , \tag{51.3}$$

where H is the strength of the external field in the "maser" coil. This trigonometric equation is solved in a graphical form. We construct a graph of the relation

$$\frac{A\tau_r}{2NHS} = \frac{1 - \cos \dfrac{A\tau}{2NHS}}{\dfrac{A\tau}{2NHS}} \, , \tag{52.3}$$

which is identical to equation (51.3) as $\tau_r = 1/2\pi\eta\gamma Q|M_p|$. This graph is given in Fig. 19a.3. From it we can conclude that excitation of the "maser" is possible only with the condition $\pi\eta\gamma Q|M_p|\tau > 1$ and that with given parameters of the "maser" coil and strength of the external field, the amplitude of the oscillations corresponds to the value $\tau = 3.5 \, \tau_r$. With a decrease in τ (an increase in q), the signal amplitude falls and tends to zero when $\tau = 2\tau_r$, while with an increase in τ it falls, tending to zero when $\tau \to \infty \, (q \to 0)$.

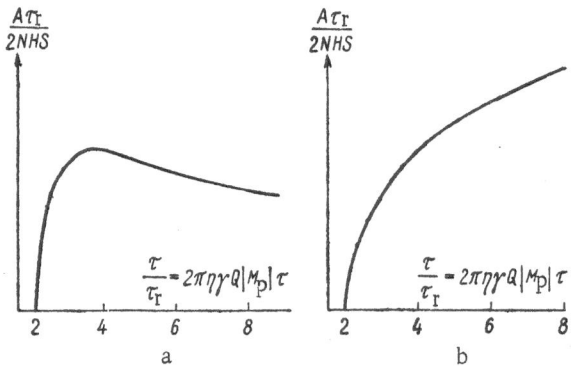

Fig. 19.3. Theoretical relation of the amplitude of the oscillations of a nuclear resonance "maser" to the parameters of the detector with a high relaxation time and liquid flow rate: a) relation to $\tau = v/q$; b) relation to the parameters of the coil.

To determine the relation of the amplitude of the oscillations to the parameters of the coil we constructed the graph given in Fig. 19b.3. This graph shows that signal amplitude increases smoothly with an increase in η, N, and Q.

Let us examine another case, when it is impossible to neglect the transverse relaxation time T_2 in the detector of the "maser." (T_2 is the effective "nutation" relaxation time.) In this case the conditions $\gamma H_1 T_2 \ll 1$; $\tau/T_2 \gg 1$ hold and the signal amplitude is described by expression (12.3). If the amplitude of the oscillations in the detector is sufficiently great, the saturation factor $Z \approx 1/\gamma^2 H_1^2 T_1 T_2 \ll 1$ and this expression has the form

$$A = \frac{4\pi\eta NSQ\omega \, |M_P|}{\gamma H_1 \tau} (1 - e^{-\tau\gamma^2 H_1^2 T_2}). \tag{53.3}$$

By substituting $H_1 = A/2N\omega S$, we obtain

$$A^2 = \frac{8\pi\eta N^2 S^2 Q\omega H \, |M_p|}{\tau} (1 - e^{-\frac{\tau T_2 A^2}{4N^2 H^2 S^2}}). \tag{54.3}$$

This transcendental equation, like the previous one, is solved graphically. Figure 20.3 gives the relation

$$\frac{A^2 \tau_r \tau}{4N^2 H^2 S^2} = 1 - e^{-\frac{\tau A^2 T_2}{4N^2 H^2 S^2}}.$$

The graph shows that excitation of the "maser" is possible on condition that $2\pi\eta\gamma Q|M_p|T_2 > 1$ and that the amplitude of the oscillations with $T_2/\tau_r > 3$ increases in proportion HSN $\sqrt{\eta QM_pT_2/\tau}$. The curve in Fig. 20b.3 gives the relation of the signal amplitude to τ, Q, η, $|M_p|$, H, N, S.

The first nuclear resonance "maser" with a flowing liquid was constructed by the French scientist Benoit [81]. In his apparatus, water was polarized by passing through a magnetic field with a strength $H_p = 7500$ oe and then it passed through a rotation coil, where the over-all magnetic moment of the protons was made negative by fast adiabatic passage through resonance. Then with a magnetization $M_p = -1.6 \cdot 10^{-6}$ erg/gaus.cc, the water passed into the coil of the "maser" circuit, which lay in the interpolar space of a magnet. The polarization, rotation, and observation of the signal were carried out with the same magnet. The nonuniformity of the external field in the coil of the "maser" circuit gave a transverse relaxation time $T_2 \approx 1.6 \cdot 10^{-3}$ sec and it was estimated from the decay time of the "wiggles" by the normal method of observing a nuclear resonance signal. The coil of the circuit, 4.5 mm in diameter and 10 mm long, contained 25 turns of wire 0.3 mm in diameter and had an inductance of 1.5 μH and a Q-factor of the order of 30.

This low Q-factor of the coil was such that the conditions for initiation of oscillations was not fulfilled and therefore it was necessary to increase the Q-factor of the coil artificially by means of the electronic circuit given in Fig. 21a.3 to Q $\approx$ 10,000. Such values could not be measured with a normal Q-meter and therefore they were estimated from the bandwidth of the circuit at a level of 0.707. The parameters of the apparatus selected guaranteed that $4\pi\eta\gamma Q$

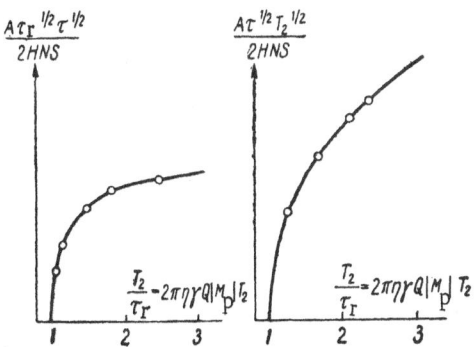

Fig. 20.3. Theoretical relation of the amplitude of the oscillations of a nuclear resonance "maser" to the parameters of the detector with a low relaxation time and liquid flow rate: a) Relation to τ, T_2, and H, S, and N; b) relation to τ, $|M_p|$, Q, η and H, S, N.

$|M_p|\, T_2 \approx 3$, and consequently oscillations were produced in it when the resonance frequency of the circuit was tuned to the precession frequency of the nuclei.

Another "maser" was constructed by Benoit in a weak magnetic field. Benzene was polarized in a strong magnetic field H_p with a strength of 7500 oe and then it flowed through a connecting tube, 2.5 m long with a diameter of 6 mm at a velocity of 96 cm/sec into a "maser" detector with a volume of 23.5 cc, lying in a magnetic field with a strength of 3.3 oe., produced by Helmholtz coils with a diameter of 32 cm. The nonuniformity of the field within the detector was 0.9 moe. The magnetization of the protons was rotated by the resonance field of a coil on the connecting tube. For increasing the relaxation time, the benzene was first boiled for a long time in a nitrogen atmosphere and the whole system was also flushed with nitrogen before filling with liquid. However, by this method it was possible to attain a relaxation time of the benzene of only 5.7 sec instead of 19 and for water it reached only 2.5 instead of 3.5 sec. The relaxation time was measured from the depolarization of the liquid in the connecting tube and therefore the low value may be explained by parasitic depolarization of the liquid in the alternating magnetic fields acting on it in flowing to the detector. The "maser" coil was 38 mm long with an internal diameter of 29.5 and an external diameter of 35 mm. It contained 930 turns of wire with a diameter of 0.3 mm and with a tuning frequency of the circuit of 14.1 kHz it had a Q-factor of 35. This Q-factor was insufficient to fulfill the conditions for initiation of the oscillations of the "maser" and therefore it was increased artificially by means of the circuit illustrated in Fig. 21b.3 to Q = 35,000. This made it possible for the "maser" to operate.

Figure 22.3 gives an oscillogram of the oscillations in the circuit. The scan was synchronized with the change in the external magnetic field. The tuning of the circuit was not changed. As the field strength approached the resonance value oscillations were produced and their amplitude increased with a change in the field, reaching a maximum value when the field strength corresponded to exact tuning of the circuit to the precession frequency of the nuclei. A further change in the field resulted in a decrease in the amplitude of the oscillations and then they stopped. By measuring the maximum amplitude of the oscillations in Fig. 22.3 and knowing the amplification factor of the receiver circuit and the oscillograph, it is possible to determine the

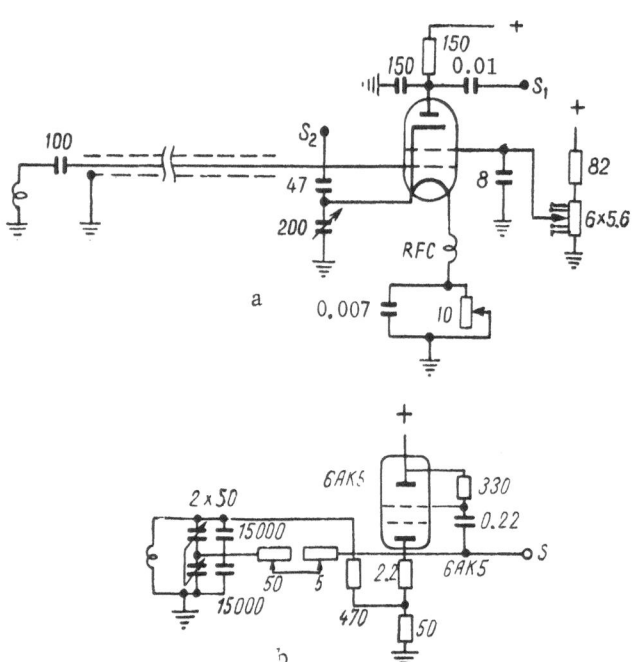

Fig. 21.3. Electronic circuits for increasing the Q-factor of the circuit of a nuclear resonance "maser;" a) in a strong magnetic field; b) in a weak magnetic field.

Fig. 22.3. Oscillogram of oscillations in the circuit of a nuclear resonance "maser."

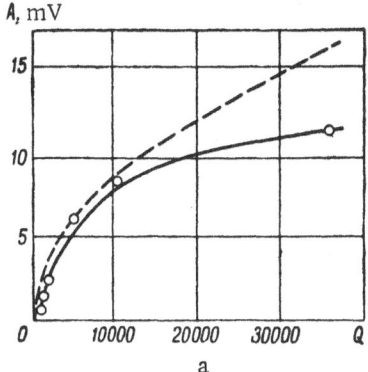

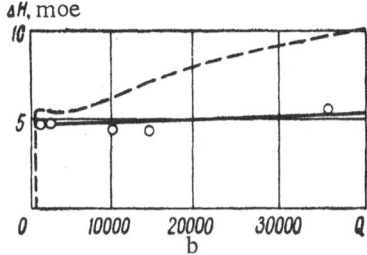

Fig. 23.3. Relation of the amplitude of the oscillations (a) and the operating bandwidth of a nuclear resonance "maser" (b) to the Q-factor of the circuit

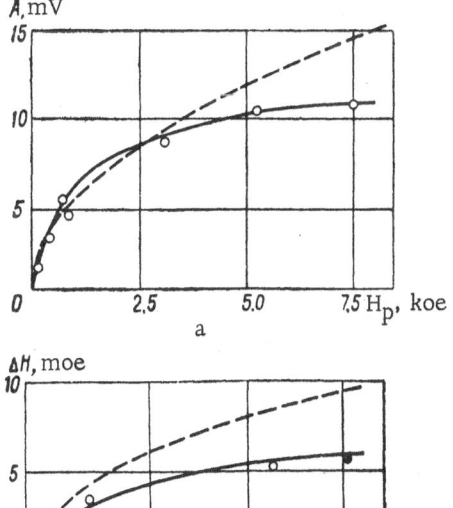

Fig. 24.3 Relation of the amplitude of the oscillations (a) and the operating bandwidth of a nuclear resonance "maser" (b) to the strength of the polarizing field.

oscillation amplitude of the "maser" at resonance A and by measuring the horizontal extent of the oscillation region and knowing the relation of the scan to the change in the field, it is possible to determine the operating bandwidth of the "maser" ΔH.

The experimental relations of A and ΔH to the Q-factor of the circuit are given in Fig. 23a and b.3. The Q-factor varied by retuning the electronic circuit used for increasing it. Values of Q below 5000 were varied through the transmission bandwidth of the circuit and values above 5000, through the time constant for attenuation of free oscillations in the circuit.

The experimental relations of A and ΔH to the strength of the polarizing field H_p, i.e., to the magnetization of the nuclei M_p, which is proportional to H_p, are given in Figs. 24a and b.3.

The relations of A and ΔH to the benzene flow rate q in the coil of the detector of the "maser" are given in Fig. 25.3. For changing the flow rate, part of the liquid was directed past the detector of the "maser" through a bypass tube; the flow rate in the tube connecting the detector to the polarizer was kept constant so that the magnetization of the nuclei in the liquid flowing into the detector remained unchanged.

The reproducibility of the experimental data was approximately 10%. The broken curves shown in Figs. 23.3-25.3 were constructed from the theoretical expressions (54.3) and (45.3) suitable for the experimental conditions. The theoretical curves in Figs. 23a.3 and 24a.3 are analogous to the curve in Fig. 20b.3. In the region of low values of Q and H_p they agree satisfactorily with the experimental points, but in the region of high values of Q and H_p, they are 30% higher. This agreement should be regarded as satisfactory as the errors in the measurement of Q, M_p, T_2, and T_1 are approximately 30%. The theoretical curve in Fig. 25a.3 shows the relation $E \sim q^{\frac{1}{2}}$ and its deviation from the experimental curve likewise does not exceed 30%. The theoretical curve in Fig. 25b.3 is a straight line as it follows from expression (45.3) that the operating range of the "maser" is independent of the liquid flow rate.

The decrease in the signal amplitude at low flow rates is explained by the fact that in the measurements it was impossible to keep the liquid flow rate in the connecting tube constant and this led to a fall in the magnetization of the nuclei M_p and, consequently, to a decrease in ΔH. The forms of the theoretical and experimental curves in Fig. 24b.3 agree quite satisfactorily. In Fig. 23b.3 there is a substantial difference in the forms of the theoretical and experimental curves which is difficult to explain.

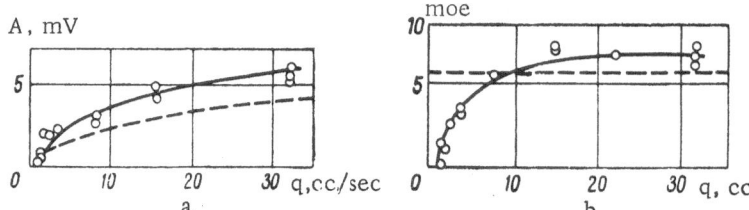

Fig. 25.3. Relation of the amplitude of the oscillations (a) and the operating bandwidth of a nuclear resonance "maser" (b) to the liquid flow rate.

TABLE 3.3

Q	450	750	2000	5000	10000	36000
Δf_{M}	2	3	6	9	12	15
Δf	2.3	3.6	7.6	12.5	16	20

Table 3.3 gives the results of an investigation of the pulling effect. The frequency shift of the "maser" oscillations Δf_{M} was measured with a fixed shift of the natural oscillation frequency of the circuit of 22 Hz. The table gives the values of Δf_{M} and also the theoretical values Δf, calculated from expression (43.3.), for various values of Q. In the calculation it was assumed that the relaxation time $T_2 = 0.086$ sec, which was estimated from the width of the nuclear resonance line, measured with the "maser" detector by the usual method of detecting the signal. The agreement is quite satisfactory. An investigation of a "maser" in a strong magnetic field was carried out by Fric [30-32, 82]. In his apparatus, tapwater with a relaxation time of 0.9 sec was pumped into a tank, from which it flowed into the interpolar space of a polarizing electromagnet. Polarization was carried out in a rubber tube 1.2 m long with a cross section of 20 mm^2, coiled into a spiral. After polarization, the liquid flowed through a tube with a cross section of 5.3 mm^2 through a rotating coil and then through the coil of the "maser" detector. Both coils lay in the field of the polarizing magnet, the detector coil in the center and the rotating coil somewhat closer to the edge so that the strength of the external field in them differed by 70 oe to avoid interaction. The rotating apparatus was described in more detail in Ch. 2. The coil of the "maser" detector was 6 mm long and contained eight turns of wire, wound on a Teflon rod 7 mm in diameter. Through the inside of the rod there passed a glass tube with an external diameter of 3.7 mm and an internal diameter of 2.6 mm. The inductance of the coil L = 0.53 mH and the quality factor Q was 76. As in previous cases, the Q-factor of the circuit was artificially increased with an electronic circuit, which consisted of an autodyne in an underexcited state.

The relation of the amplitude of the oscillations in the "maser" circuit to the Q-factor of the circuit at several liquid flow rates is given in Fig. 26a.3 and Fig. 26b.3 shows the relation of the amplitude of the oscillations to the flow rate at several values of Q, constructed from the same results. The curves have shapes that are different from those in Benoit's experiments. This is explained by the fact that in Benoit's apparatus the liquid flowed through the coil of the "maser" for a time of the order of 1 sec and therefore the relaxation process had a substantial effect, while in Fric's apparatus the time to flow through varied over the range 6-30 msec so that here the processes in the "maser" were closer to the case described by equation (51.3). This is confirmed by the similarity of the experimental curves in Fig. 26b.3 to the theoretical curve in Fig. 19a. 3. The pulling effect was investigated on the same apparatus and the change in the oscillation frequency of the "maser" Δf_{M} with a change in the natural frequency of the circuit by 100 Hz was measured for various values of Q and liquid flow rates q. This relation is shown in Fig. 27a. 3. The broken curves were constructed from

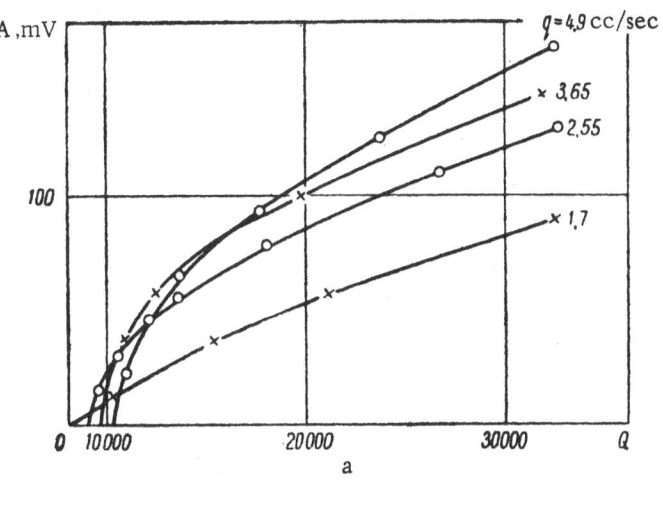

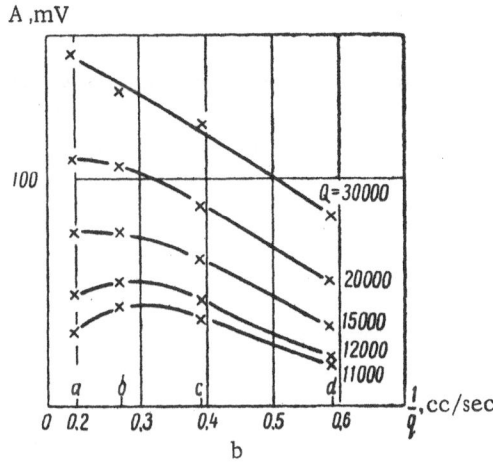

Fig. 26.3. Relation of the amplitude of the oscillations of a nuclear
resonance "maser" to the Q-factor of the circuit (a) and the liquid
flow rate (b).

the theoretical expression with the value of T_2 in each case selected to give the best agreement between the theoretical and experimental curves. It was observed that different optimal values of T_2 corresponded to curves obtained with different flow rates. The relation of T_2 to the liquid flow rate is given in Fig. 28.3.

The experimental relation of the operating range of the "maser" δf_M to the Q-factor of the circuit with a flow rate of 4.9 cc/sec is given in Fig. 29.3. The broken curve was constructed from expression (45.3) for the experimental conditions.

The relation of the pulling of the "maser" frequency in a weak field to the width of the nuclear resonance line ΔH is given in Fig. 27b.3 [34]. The shift in the oscillation frequency of the "maser" Δf_M was measured for a shift in the strength of the external field by a definite value, corresponding to a difference between the nuclear precession frequency and the natural frquency of the circuit of 100 Hz. The width of the nuclear resonance line was regulated by changing the nonuniformity of the external field in the volume of the detector and was measured from the width of the nuclear absorption signal. The square of $T_2 \approx 2/\gamma \Delta H$ is plotted along the abscissa axis in Fig. 27b.3. The experimental points lay on a straight line, which indicates a quadratic relation between the frequency shift and the transverse relaxation time of the nuclei in the "maser" coil. At first glance this is contradictory to (42.3), from which it follows that

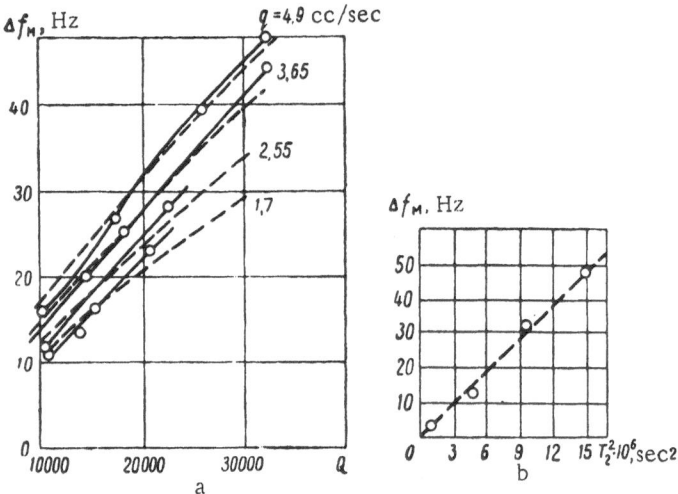

Fig. 27.3. Relation of the change in the frequency of the "maser" with a change in the natural frequency of the circuit by 100 Hz to the Q-factor of the circuit and the liquid flow rate (a) and to T_2 (b).

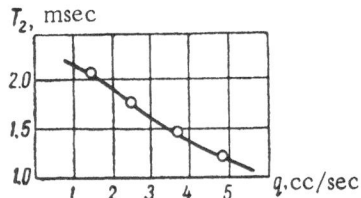

Fig. 28.3. Relation of the effective transverse relaxation time to the liquid flow rate.

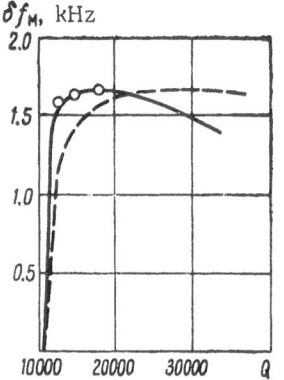

Fig. 29.3 Relation of the operating range of the "maser" to the Q-factor of the circuit.

$$\Delta f_{\text{м}} = \frac{\omega - \omega_{\text{г}}}{\omega_{\text{г}} - \omega_0} \Delta f_0 = \frac{\omega T_2}{\omega T_2 + 2Q} \Delta f_0. \tag{55.3}$$

The relation obtained may be explained by the fact that the same amplitude of the "maser" oscillations A_M was maintained in the experiment at all values of ΔH. As follows from relation (54.3), which is valid for a "maser" in a weak field with a long residence time of the nuclei in the coil, this is guaranteed in conditon that $2\pi\gamma Q \mid M_p \mid = \text{const}$. In practice, for maintaining a constant oscillation amplitude it was necessary to change the degree of positive feedback of the circuit, i.e., to set the A-factor in proportion to the nonuniformity of the field in the circuit

$$Q \sim \frac{1}{T_2}.$$

66

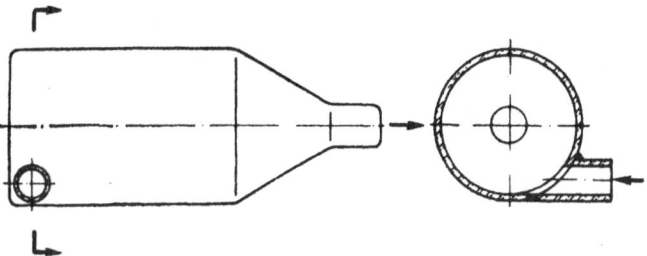

Fig. 30.3 Diagram of flow detector with rotating liquid.

Allowing for the fact that in a sufficiently nonuniform field $\omega T_2 \ll Q$, from expression (55.3) it follows that there is a quadratic relation between Δf_M and T_2.

The pulling effect is the main drawback of "masers" used for the measurement and stabilization of magnetic fields. The difference between the frequency of the "maser" ω and the nuclear precession frequency may be determined from expression (43.9) as

$$\omega - \omega_0 = \frac{2Q}{\omega T_2 + 2Q}\,(\omega_r - \omega_0),$$

where ω_r is the natural frequency of the circuit. To bring ω closer to ω_0 it is necessary to use "masers" with a low value of Q and a high value of T_2.

The value of T_2 is limited by the nonuniformity of the external field. The maximum value $T_2 = 0.3$ sec was obtained under laboratory conditions with the detector illustrated in Fig. 30.3 with a rotating flowing liquid. An iron-clad electromagnet (Ch. 1) giving a very low stray field with a polarizing field strength H_p = 20,000 oe was used. The high values of T_2 and M_p made it possible to achieve autooscillation of the "maser" with a Q-factor of the coil of 26 without artificially increasing it in a magnetic field with a strength of the order of 1 oe. In this case $(\omega - \omega_0)/(\omega_r - \omega_0) = 1/75$ and with a decrease in the nonuniformity of the field, this ratio must fall quadratically.

CHAPTER 4

INSTRUMENT AND RADIATION EFFECTS

1.4. Instrument Effect

In experiments on the nutation of the magnetization of the nuelci in a flowing liquid by a resonance oscillating field (Ch. 2), it was noted that maximum nutation was observed in some nutation detector designs at an oscillating field frequency which differed from the nuclear precession frequency by a certain value, which was proportional to the liquid flow rate. This effect was called the "instrument effect" [34].

Theory of the Effect. In observing nuclear magnetic resonance it is usual to use a linearly polarized oscillating magnetic field, which is equivalent to the sum of two fields with circular polarization, rotating in opposite directions. The active field is the one rotating in the direction of precession of the nuclei and is the one which induces resonance, while the second field has no substantial effect on the nucleus and only produces a slight shift of the precession frequency [95-97]. A nuclear resonance effect is observed if the active component of the oscillating field acting on the nucleus rotates with the nuclear precession frequency ω_0. If the nucleus is stationary, the frequency of the rotating components acting on it coincides with the frequency of the oscillating field and resonance is observed when the frequency of this field $\omega = \omega_0$. If the nucleus is moving, then the frequency of rotation of the rotating components acting on it may not coincide with the frequency of the oscillating field and nuclear resonance is observed when $\omega \neq \omega_0$.

Let us introduce a system of Cartesian coordinates with the origin at the geometric center of the detector coil. Let the z axis lie parallel to the external field and the x and y axes perpendicular to it. The component of the oscillating field which is perpendicular to the external field participates in the resonance effect and therefore, we will consider its projection H_1 on the plane xy, which varies in accordance with the equation $H_1 = H_{10} \cos \omega t$. If the direction of the field H_1 at some point in the detector makes an angle α with the x axis, then the projections of H_1 on the coordinate axes will be

$$\left.\begin{aligned} H_{1x} &= H_{10} \cos \omega t \cdot \cos \alpha, \\ H_{1y} &= H_{10} \cos \omega t \cdot \sin \alpha. \end{aligned}\right\} \tag{1.4}$$

The oscillating field is equivalent to the sum of two fields of half the amplitude (H_1' and H_1'') rotating in opposite directions. The projections of the vectors of these rotating fields on the coordinate axes will be

$$\left.\begin{aligned} H_{1x}' &= \frac{H_{10}}{2} \cos (\omega t + \alpha), & H_{1y}' &= \frac{H_{10}}{2} \sin (\omega t + \alpha), \\ H_{1x}'' &= \frac{H_{10}}{2} \cos (\omega t - \alpha), & H_{1y}'' &= \frac{H_{10}}{2} \sin (\omega t - \alpha). \end{aligned}\right\} \tag{2.4}$$

These expressions show that the angular rotation frequency of the fields H_1' and H_1'' equals $\omega + d\alpha/dt$ and $\omega - d\alpha/dt$ respectively. The angle α is constant at every point of the detector and therefore $d\alpha/dt = 0$, i.e., a rotating field with a frequency equal to the frequency of the oscillating field ω acts on a stationary nucleus. A rotating field with a frequency $\omega \pm d\alpha/dt$ acts on a moving nucleus. The value $d\alpha/dt$ determines the displacement of the resonance frequency from the nuclear precession frequency in the external field. The magnitude of the displacement depends on the velocity and direction of movement of the nuclei and the form of the lines of force of the oscillating field and the sign of the displacement is determined by the sign of γ. Thus, in the observation of the nuclear resonance of moving nuclei, the resonance frequency of the oscillating field at each point in the detector with the coordinates x, y, and z differs from the nuclear precession frequency at this point by the value $\Delta\omega_A = d\alpha/dt\,(x, y, z)$. This frequency shift produces a shift of the nuclear resonance line by $\Delta\omega_A$, averaged through the volume of the detector v:

$$\Delta\overline{\omega}_A = \frac{1}{V} \int\int\int_V \frac{d\alpha}{dt}\,dv. \tag{3.4}$$

If $\Delta\omega_A$ is not the same through the whole volume of the detector, then there is broadening of the nuclear resonance line. In this case the width of the line

$$\delta\omega_A = \Delta\omega_{A\,\text{max}} - \Delta\omega_{A\,\text{min}}, \tag{4.4}$$

where $\Delta\omega_{A\,\text{max}}$ and $\Delta\omega_{A\,\text{min}}$ are the maximum and minimum values of $\Delta\omega_A$ in the volume of the detector.

If the detector coil and the trajectory of the nuclei lie symmetrically relative to a plane parallel to the external field, then $\Delta\overline{\omega}_A = 0$, which means that there is no shift of the resonance line. In this case $|\Delta\omega_{A\,\text{max}}| = |\Delta\omega_{A\,\text{min}}|$ and the broadening of the line is expressed by

$$\delta\omega_A = 2\left(\frac{d\alpha}{dt}\right)_{\text{max}}. \tag{5.4}$$

By means of expressions (3.4)-(5.4) it is possible to determine $\Delta\overline{\omega}_A$ and $\delta\overline{\omega}_A$ theoretically for each actual detector design.

Experimental Investigation of the Effect. The maximum shift of the resonance line caused by the instrument effect is of the order of a few hundredths of an oersted. In order to observe this shift clearly it should be investigated using an external field with a strength comparable in magnitude and therefore the instrument shift of the nutation signal was observed experimentally in a field with a strength H = 0.07 oe. The experimental equipment consisted of an iron-clad polarizing magnet and an absorption detector with a volume of 0.03 cc (see Fig. 3a.3). The strength of the polarizing field H_p = 5000 oe. The absorption signal was observed in a uniform field of 20 oe with an autodyne detector. The connecting tube had a diameter of 0.3 cm.

For measuring the instrument shift, the nutation detector investigated was placed in a weak field and connected into the tube between the polarizing field and the absorption detector. Its coil was connected to the output of a ZG-10 audiofrequency oscillator. The first nutation extremum was produced by changing the frequency and potential at the output of the oscillator. The resonance frequency of the oscillating field was difficult to determine directly as quite a large shift in the frequency of the oscillating field had little effect on the amplitude of the signal because of its great width. It was possible to establish much more accurately the frequencies at which inversion of the absorption signal occurred, i.e., it vanished. By determining both these frequencies from the dial of the oscillator it was possible to find the resonance frequency as the arithmetic mean of these two values. When the resonance frequency ω_1 had been determined in this way, the connecting tubes were changed over so that the liquid flowed through the nutation detector in the opposite direction and the resonance frequency ω_2 was determined.

According to expression (3.4), the instrument frequency shift is proportional to $d\alpha/dt$, averaged through the volume of the detector. If x is the coordinate of a nucleus measured along its path and W is the velocity of the nucleus, then

$$\frac{d\alpha}{dt} = \frac{d\alpha}{dx}\,W. \tag{6.4}$$

69

All the working spaces investigated had inlets identical to the outlets so that a change in the direction of the liquid flow did not change the trajectory of the nuclei or W, but only the sign of the velocity of the nuclei. Under these conditions, the instrument frequency shift obviously changed only in sign, i.e.,

$$\Delta \omega_A = \left| \frac{\omega_1 - \omega_2}{2} \right| .$$

The second method of observing the instrument effect consisted of measuring the resonance frequency ω_1, rotating the nutation detector through 180° about an axis perpendicular to the external field or reversing the direction of the external field, and then measuring the resonance frequency ω_2. A change in the relative direction of the z axis and the vector of the external field changes the direction of nuclear precession relative to the coordinate axes, i.e., it changes the active rotating component of the oscillating field. Obviously there is also a change in the sign of $d\alpha/dt$ and the sign of the instrument frequency shift, i.e.,

$$\Delta \bar{\omega}_A = \left| \frac{\omega_1 - \omega_2}{2} \right| .$$

These two methods were used to investigate several designs of nutation detectors. Fig. 1.4 shows schematically the designs in which an instrument shift was observed and Fig. 2.4 shows designs with which there was no instrument shift. In the detectors shown in Fig. 2.4, the trajectory of the nuclei and the coils are symmetrically placed relative to one of the planes parallel to the external field (this plane passes vertically down the axis of the detector and is a plane of symmetry of the coil). In the detectors illustrated in Fig. 1.4 there is no such

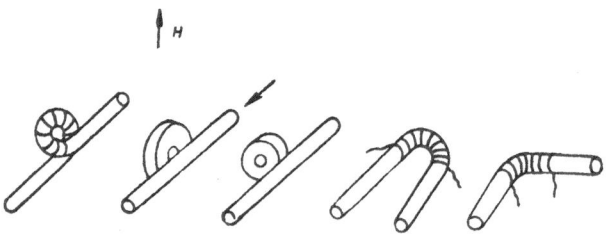

Fig. 1.4. Designs of nuclear resonance flow detectors in which an instrument frequency shift was observed.

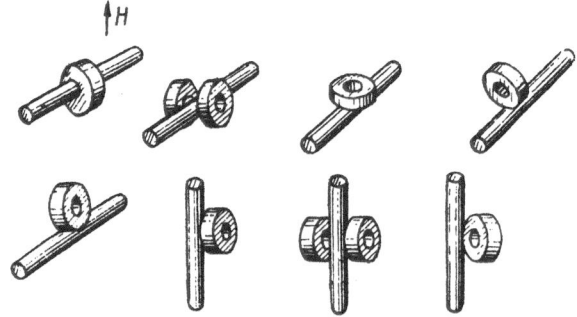

Fig. 2.4. Designs of nuclear resonance flow detectors in which no instrument frequency shift was observed.

70

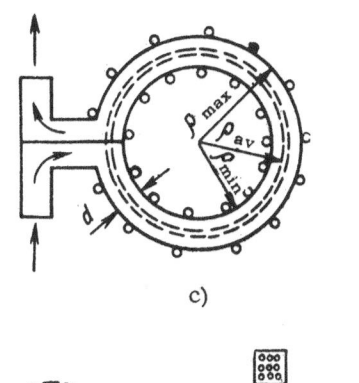

c)

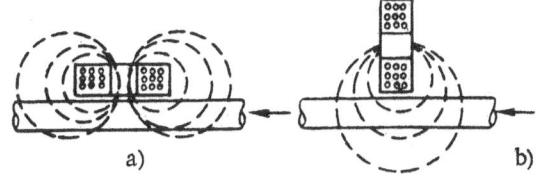

a)　　　　　　　　b)

Fig. 3.4. Schematic cross sections of detectors of various designs: a, b) detectors with flat coils; c) toroidal detector.

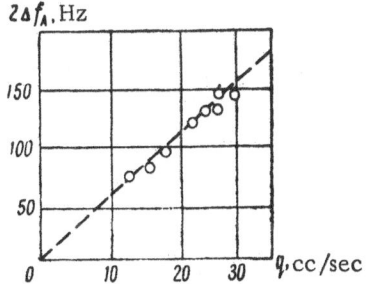

Fig. 4.4 Relation of the instrument shift of the nuclear magnetic resonance frequency in a toroidal detector to the liquid flow rate.

plane. Thus, according to theory, if there is a common plane of symmetry of the coil and the liquid stream, parallel to the external field, the nutation detector does not show an instrument shift. If there is no such plane, an instrument shift of the resonance frequency is observed.

Fig. 3.4 (a and b)shows the cross sections of two of the detectors illustrated in Fig. 1.4. The broken lines show the lines of force of the oscillating field. When the liquid flows from right to left, the lines of force of the oscillating field acting on the nuclei rotate clockwise and when the external field is directed away from us, there must be a negative instrument frequency shift. This was observed experimentally.

The cross section of a toroidal detector similar to that illustrated in Fig. 1.4 is given in Fig. 3c.4. The broken lines show the lines of force of the oscillating field. This figure shows that when the liquid flows the lines of force of the oscillating field acting on the nuclei rotate anticlockwise and $d\alpha/dt = 1/\rho$, i.e., from expression (4.4) $d\alpha/dt = W/\rho$. By substituting this value in expression (3.4) and integrating, we obtain

$$\Delta\bar{\omega}_A = \frac{W}{d}\ln\frac{\varrho_{max}}{\varrho_{min}} \approx \frac{W}{\varrho_{av}}\left(1 + \frac{d^2}{12\varrho_{av}^2} + \frac{d^4}{80\varrho_{av}^4}\right), \qquad (7.4)$$

where ρ_{max} and ρ_{min} are the maximum and minimum radii of the torus, d is the diameter of a section of the torus, and ρ_{av} is the mean radius of the trajectory of the nuclei.

Fig. 4.4 gives the experimental relation of the shift in the nuclear resonance frequency with a change in the direction of the liquid flow in a toroidal detector to the liquid flow rate. The mean radius of curvature of the torus was 0.8 cm and the internal diameter of the tube d = 0.3 cm. The broken line in Fig. 4.4 represents the theoretical relation (7.4) of the frequency shift to the flow rate, which has the following form after the substitution of the values of ρ_{av} and d:

$$2\Delta f_A = 5.6q.$$

The instrument broadening of nuclear resonance is of the same order as that produced by the finite time it takes for the nuclei to pass through the detector coil and therefore it is possible to determine experimentally only the total width of the line produced by these effects, which, as was shown previously, in a cylindrical flow detector equals(1/8) q/v Hz.

It should be noted that in some cases the instrument effect may produce a shift in the nuclear resonance signal in detectors of the designs shown in Fig. 2.4. This may occur when the nuclei passing through the detector on different sides of the plane of symmetry of the flow and the coil mentioned above make different contributions to the nuclear resonance effect. The sign of the instrument effect is determined by the direction of bending of the lines of force of the oscillating field in the region of the most intense nuclear resonance effect. The dissymmetry may be produced by careless construction of the detector and also by the nonuniformity of the external field, distributed irregularly through the volume of the stream in the detector.

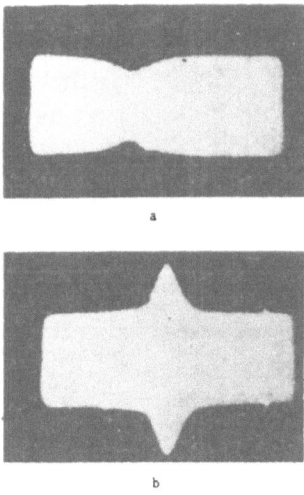

Fig. 5.4. Photograph of nuclear magnetic resonance
signals: a) absorption signal; b) emission signal.

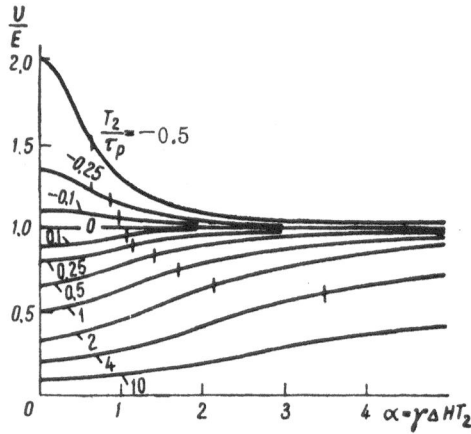

Fig. 6.4. Theoretical relation of the nuclear resonance signal amplitude to the frequency difference
at various ratios T_2/τ_r.

2.4. Radiation Damping

Benoit investigated the radiation damping effect in a flow detector [33]. He obtained the oscillograms
of the nuclear resonance signal shown in Fig. 5.4. The potential from the coil of the nuclear resonance detector circuit was fed to the vertical plates of the oscillograph. A signal was observed from protons in a field with
a strength of about 1 oe and therefore the frequency of the potential on the coil was approximately 4 kHz.
The horizontal scan was synchronized with the slow sawtooth modulation of the magnetic field. At the moment
when the field strength corresponded to the resonance value there was the maximum change in the potential on
the coil of the circuit   An oscillogram was obtained with positive magnetization (see Fig. 5a.4) of the protons
M_0 in a liquid (benzine) flowing into the detector from the polarizing field, i.e., it represents a nuclear absorption signal. Another oscillogram (see Fig. 5b.4) was obtained with negative magnetization of the protons in the
liquid flowing into the detector (reversal was achieved by rapid adiabatic passage through resonance in the tube
between the polarizing field and the nuclear resonance detector). This oscillogram represents a nuclear emission signal.

It can be seen readily that despite the completely identical conditions the emission signal has a greater amplitude than the absorption signal. This effect is explained by radiation damping of the nuclei in the circuit.

A circuit similar to Rollin's circuit was used for detection of the signal. The change ΔE in the potential E in the coil of the circuit at the onset of conditions for nuclear resonance for preliminarily polarized protons of the liquid passing through it was observed. The coil was inductively coupled with the audiofrequency generator inducing the emf E in it. The current I in the coil may be found from the expression

$$E + I\left(\,i\,L\omega + r + \frac{1}{i\,C_0\omega}\right) = 0, \tag{8.4}$$

where L, C_0, and r are the inductance, capacitance, and active resistance of the circuit:

$$L = L_0\left[1 + 4\pi\eta\left(X' - iX''\right)\right]. \tag{9.4}$$

Here L_0 is the inductance of the coil in the absence of conditions for nuclear resonance, η is the filling coefficient of the coil with the liquid, and X' and X" are the dynamic magnetic susceptibilities of the nuclei.

The potential in the coil U = L ωI may be found from the equation

$$\frac{E}{U} + 2\left[\frac{\omega - \omega_p}{\omega_p} + 4\pi\eta X'\right] - i\left[\frac{1}{Q} + 4\pi\eta X''\right] = Q, \tag{10.4}$$

where $\omega_p = 1/\sqrt{L_0 C_0}$; and Q is the quality factor of the circuit. The solution of this equation for the condition when X' and X" are described by the expressions (35.3) and (36.3) was found for the case when $\omega = \omega_r$, i.e., when the oscillator frequency equals the natural oscillation frequency of the circuit, while the external magnetic field varies close to the resonance value $H_0 = \omega_r/\gamma$. This gave

$$|U| = |E|\,Q\left[\left(\frac{T_2}{\tau_r}\cdot\frac{\alpha}{1+\alpha^2}\right)^2 + \left(1 + \frac{T_2}{\tau_r}\cdot\frac{1}{1+\alpha^2}\right)^{-\frac{1}{2}}\right], \tag{11.4}$$

where $\alpha = \gamma(H - H_0)T_2$; T_2 is the transverse relaxation time, $\tau_p = 1/2\pi\eta\gamma QM$ is the radiation relaxation time, and M is the magnetization of the nuclei in the detector. This relation is given in Fig. 6.4. The negative values of T_2/τ_r correspond M < 0. The amplitude of the nuclear resonance signal is the deviation of each curve from the line U/E = 1. The vertical line on each curve marks the position of the half-width of the signal at half-height. On examining Fig. 6.4 it is readily seen that with the same absolute value of T_2/τ_r the nuclear resonance signal has a greater amplitude when $T_2/\tau_p < 0$, than when $T_2/\tau_r > 0$. This corresponds to the experimental result. The theoretical ratio of the amplitudes of the emission and absorption signals equals $(1 + T_2/\tau_r)/(1 - T_2/\tau_r)$, while the relative increase in the width of the signal is $\frac{3}{4}\,T_2/\tau_r - \frac{3}{32}\,(T_2/\tau_r)^2$.

Part II

Application of Nuclear Magnetic
Resonance in a Flowing Liquid

CHAPTER 5

MEASUREMENT AND STABILIZATION OF WEAK MAGNETIC FIELDS

1.5. Comparison of Normal and Flow Nuclear Resonance Detectors

Since the first work on nuclear resonance there have appeared many instruments which make it possible to use this method for the absolute measurement [98-106] and stabilization [107-111] of magnetic fields. The advantage of these instruments is their high accuracy and the drawback, the fact that it is impossible to use them in fields below 300 oe. Only in a few experiments has it been possible to use nuclear resonance in weaker fields.

For example, fields down to 150 oe were measured with a detector with a volume of 9 cc [112] and fields down to 40 oe were measured with a 5-cc detector [59]. The stabilization of fields with a strength down to 120 oe with an accuracy of $2 \cdot 10^{-5}$ has also been achieved [113]. To obtain signals in weaker fields [114, 115] it is necessary to have a very large detector volume and a very uniform field. Thus, a proton resonance signal was obtained in fields of 6-12 oe with a detector with a volume of 1000 cc [116, 117] and in fields of 3-0.5 oe with a detector with a volume of 2000 cc [118, 119] with a nonuniformity of the field in the detectors no higher than 10^{-4} oe/cm. The free precession method suffers from the same drawbacks [120-127] and therefore it is used only in geomagnetometers. Thus, with a detector volume of 500 cc the nonuniformity of the field measured should not exceed $1.2 \cdot 10^{-4}$ oe/cm [126]. Let us examine the reasons for the above drawback of the method.

In a detector with a stationary substance, the amplitude of the nuclear resonance signal in a uniform field with the optimal strength of the oscillating field is determined by the following expression:

$$A = 4\pi\eta NSQX_0\gamma H^2 \frac{T_2^*}{T_1} , \qquad (1.5)$$

where η, N, S, and Q are the filling factor, the number of turns, the cross section, and the quality factor of the coil of the detector circuit, X_0 is the static magnetic susceptibility of the nuclei in the substance, γ is the gyromagnetic ratio of the nuclei, H is the strength of the external magnetic field in the detector, and T_1 and T_2^* are the effective relaxation times of the components of the magnetization of the nuclei along and across the external field in the detector.

The quadratic relation of the signal amplitude to the field strength makes it impossible to use a normal nuclear resonance detector in practice for measuring and stabilizing weak magnetic fields as the only possibility of obtaining a satisfactory signal-to-noise ratio in a weak field lies in increasing the detector volume to compensate for the decrease in H with an increase in S.

The expression for the amplitude of a nuclear resonance signal in a flow detector was obtained in Ch. 3. Under optimal conditions it has the form:

$$A = 2.8\pi\eta NSQX_0\gamma HH_p, \qquad (2.5)$$

where H_p is the strength of the polarizing field. A comparison of expressions (1.5) and (2.5) shows that with the same external field H, the amplitude of the signal in a flow detector exceeds that of the signal in a normal detector by a factor of at least H_p/H as T_2^* is always less than T_1. For example, in a field with a strength H = 1 oe with a polarizing field H_p = 10,000 oe, the amplitude of the signal in a flow detector must exceed that in a normal detector by 10,000.

Let us find the ratio of the volumes of normal and flow detectors which will give the same signal amplitudes. For this purpose we equate expressions (1.4) and (2.4), denoting the parameters of the normal detector by the subscript 0 and those of the flow detector by the subscript a. After cancelling common factors, we obtain

$$2Q_0 N_0 S_0 H \frac{T_{20}^*}{T_{10}} = 1.4 Q_a N_a S_a H_{\mathrm{p}}. \qquad (3.5)$$

As $S \sim d^2$, where d is the diameter of the detector, with an inductance constant $N \sim d^{-1/2}$ and $Q \sim d$, after rearranging equation (3.5), we obtain

$$\frac{d_0}{d_a} \sim \left(\frac{H_{\mathrm{p}}}{H} \right)^{2/5} \left(\frac{T_{10}}{T_{20}^*} \right)^{2/5}. \qquad (4.5)$$

The volume of a normal detector $v_0 \sim d_0^3$. The volume of a flow detector $v_a \sim d_a^3$. Then from expression (4.5) we find:

$$\frac{v_0}{v_a} = \left(\frac{H_{\mathrm{p}}}{H} \right)^{6/5} \left(\frac{T_{10}}{T_{20}^*} \right)^{6/5}. \qquad (5.5)$$

As $T_{20}^* \ll T_{10}$ and $H_{\mathrm{p}} \gg H$, then $d_0 \gg d_a$ and $v_0 \gg v_a$, i.e., to obtain a good signal in a weak field the size of the flow detector required is much less than that of the normal detector.

In practice, with a flow detector a signal-to-noise ratio greater than 10 was obtained in a field of 4 oe and above with the volume $v_a = 0.03$ cc, i.e., less than the volume of a normal detector in such fields by a factor of 10^4 - 10^5. Moreover, with a flow detector it is possible to obtain a nuclear resonance signal in weak fields with a nonuniformity greater than 10 oe/cm, i.e., 10^4-10^5 times greater than with a normal detector (Ch. 3). Thus, the use of a flow detector makes it possible to use nuclear magnetic resonance for measuring and stabilizing weak magnetic fields with considerable nonuniformity.

The drawback of preliminary polarization of a flowing liquid is the presence of a strong polarizing field, which may affect somewhat the strength of the field measured. For this reason it is most profitable to use this method for measuring and stabilizing fields in cases where it is important to establish and maintain a definite field strength regardless of the methods used to produce the field. When an iron-clad magnet or a specially designed system of Helmholtz coils is used for polarization, a flow detector may be used for absolute measurements of a magnetic field with high accuracy.

2.5. Measurement Errors

For measuring a magnetic field by the nuclear resonance method it is necessary to change the frequency of the oscillating field smoothly and determine the moment when this frequency exactly equals the precession frequency of the nuclei f_0 in the detector. The relation of the potential U at the output of the phase detector, connected to the nuclear resonance detector circuit, to the frequency f, close to the value f_0, is given in Fig. 1.5. When the frequency approaches f_0 there appears a nuclear resonance signal in the form of an alternating potential with a frequency equal to the modulation frequency of the field measured f_m. At the output of the phase detector there appears a positive potential, whose magnetude is proportional to the signal amplitude. At some value $f = f_1$, the signal amplitude and, consequently, the potential at the output of the phase detector have a maximum and then they begin to fall. When $f = f_0$, the amplitude of the signal with a frequency f_m and the potential at the output of the phase detector equal zero. With a further increase in f, when $f > f_0$, the signal appears again, but in antiphase, and in the phase detector there appears a negative potential proportional to the signal amplitude. If the frequency $f = f_2$, the signal amplitude again has a maximum, to which the minimum potential at the output of the phase detector corresponds. When $f > f_2$, the amplitude of the signal falls smoothly to zero.

Thus, to measure a magnetic field it is sufficient to determine the frequency $f = f_0$ at which $U = 0$. In practice it is impossible to fix exactly the frequency at which $U = 0$. It is only possible to determine a certain frequency region $f_0 - \Delta f < f < f_0 + \Delta f$, in which U is less than the noise level. The half-width of this region Δf determines the error of the measurement of the field strength $\Delta H = 2\pi \Delta f/\gamma$. The same value determines the error of the stabilization of the field as with a change in the field of $\pm \Delta H$, no detuning signal that can be distinguished on the noise background appears at the output of the phase detector. The value Δf depends on the noise amplitude U_{no} and the derivative $\partial U/\partial f$ at $U = 0$:

$$\Delta f = \frac{U_{no}}{\left(\dfrac{\partial U}{\partial f}\right)_{U=0}} . \tag{6.5}$$

For an estimate it may be assumed that

$$\left(\frac{\partial U}{\partial f}\right)_{U=0} = \frac{2U_{max}}{\delta f} , \tag{7.5}$$

where U_{max} is the maximum amplitude of the nuclear resonance signal, i.e., the absolute value of the potential at the output of the phase detector when f and $f_1 = f_2$, and $\delta f = f_2 - f_1$ is the width of the nuclear resonance signal frequency units. Then

$$\Delta f = \frac{U_{no}\delta f}{2U_{max}} = \frac{\delta f}{2a} , \tag{8.5}$$

or

$$\Delta H = \frac{\delta H}{2a} , \tag{9.5}$$

where $a = U_{max}/U_{no}$ is the signal-to-noise ratio and $\delta H = 2\pi \delta f/\gamma$ is the signal width in units of field strength.

The width of the nuclear resonance signal is determined by the following factors: the nonuniformity of the external field, the strength of the oscillating field, and the broadening proportional to the liquid flow rate q and the amplitude and frequency of the modulation of the field measured.

With a modulation amplitude greater than the half-width of the resonance line there is broadening of the signal, while with an amplitude less than the half-width of the resonance line, the signal amplitude falls. The optimal modulation amplitude is of the order of the half-width of the nuclear resonance line. It should guarantee the maximum value of $dU/\partial f$ when $U = 0$. The field modulation frequency f_m should satisfy the condition $f_m \ll \delta f$. In this case the broadening associated with the modulation frequency is negligibly small.

In a cylindrical detector with the optimal modulation amplitude and the optimal amplitude of the oscillating field, if the nonuniformity of the field measured is negligibly small, i.e.,

$$\operatorname{grad} H \ll \frac{q}{\gamma v_a d_a} ,$$

where v_a is the volume of the detector and d_a is the diameter of the detector, then the width of the nuclear resonance signal is determined by the expression

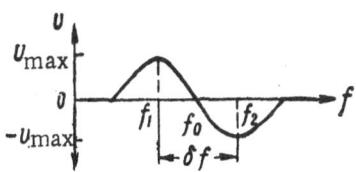

Fig. 1.5. Form of nuclear absorption signal with slow transit.

$$\delta H = \frac{5q}{\gamma v_a} \ . \tag{10.5}$$

If the amplitude of the oscillating field is less than the optimal value, as is usually the case when autodyne nuclear resonance detectors are used, then

$$\delta H = \frac{2\pi q}{8\gamma v_a} \ . \tag{11.5}$$

By substituting these values in expression (9.5) we obtain the relation of the measurement error to the flow rate: when

$$H_1 = H_{1opt} \ , \qquad \Delta H = \frac{2.6q}{\gamma v_a a} \ ; \tag{12.5}$$

when

$$H_1 \ll H_{1opt} \ , \qquad \Delta H = \frac{\pi}{8} \cdot \frac{q}{\gamma v_a a} \ . \tag{13.5}$$

The signal amplitude and, consequently, the signal-to-noise ratio are porportional to $A_M = -4\pi \eta NSHQ \gamma M_p$, which is proportional to $d_a^{5/2}$ as the area of the coil $S \sim d_a^2$, with a constant inductance of the coil, the number of turns $N_a \sim d_a^{-1/2}$, and the quality factor $Q \sim d_a$. Consequently, $a \sim d_a^{5/2}$, while $v_a \sim d_a^2 l_a$, where l_a is the length of the detector. By substituting these values in expressions (12.5) and (13.5) we obtain:

$$\Delta H \sim \frac{q}{d_a^{9/2} l_a} \ . \tag{14.5}$$

In a nonuniform field, if $H \gg q/\gamma v_a d_a$, the width of the signal is determined by the nonuniformity of the field in the detector and

$$\delta H = \frac{d_a \, \mathrm{grad} \, H}{2} \ . \tag{15.5}$$

By substituting this value in expression (9.4) we obtain the measurement error

$$\Delta H = \frac{\mathrm{grad} \, H d_a}{4a} \ . \tag{16.5}$$

As was shown in Ch. 3, in this case the signal amplitude is determined by the expression

$$A = \frac{4A_M q}{v_a d_a \gamma \, \mathrm{grad} \, H} \ ,$$

whence

$$a \sim \frac{q}{d_a^{1/2} l_a}$$

and the measurement error

$$\Delta H \sim \frac{d_a^{3/2} l_a}{q} \ . \tag{17.5}$$

Thus, with a low flow rate, when $q \ll v_a d_a \gamma \operatorname{grad} H$, the measurement error falls with an increase in q and with a high flow rate, when $q \gg v_a d_a \gamma \operatorname{grad} H$, the error falls with a decrease in q; consequently, the minimal measurement error corresponds to some optimal flow rate

$$q_{opt} \approx v_a d_a \gamma / \operatorname{grad} H. \tag{18.5}$$

In this case the contribution to the width of the line proportional to the liquid flow rate and the contribution introduced by the nonuniformity of the field are comparable in magnitude.

3.5. Optimal Parameters of Apparatus

The role of the polarizing device consists of ensuring that there enters the nuclear resonance detector liquid with the highest possible magnetization of the nuclei. On this basis we can select the strength of the polarizing field H_p, the polarizing volume v_p, and the volume of the connecting tube v_t. From expression (6.1) it follows that for the magnetization of the nuclei with given values of v_p and v_t it is possible to obtain the optimal liquid flow rate

$$q_{p\,opt} = \frac{v_p}{T_1 \ln \frac{v_p + v_t}{v_t}}, \tag{19.5}$$

where T_1 is the natural spin-lattice relaxation time at which the magnetization is maximal and given by

$$M_{max} = X_0 H_p \frac{v_p}{v_p + v_t} \left(\frac{v_t}{v_p + v_t} \right)^{\frac{v_t}{v_p}}. \tag{20.5}$$

Let us introduce the value $K = v_t/v_p$ and then expression (20.5) assumes the form

$$M_{max} = X_0 H_p \frac{K^K}{(1+K)^{1+K}}. \tag{21.5}$$

The relation of $M_{max}/X_0 H_p$ to K is shown in Fig. 2.5.

This relation shows that when $K = 0.1$ the magnetization of the nuclei is 30% less than the maximum, while when $K = 0.5$ it equals only 38% of the maximum value, i.e., $K < 0.1$ should be chosen.

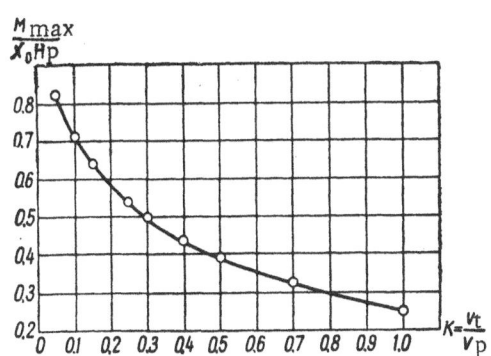

Fig. 2.5. Theoretical relation of the signal amplitude to the ratio of the volume of the connecting tube v_t to the volume of the polarizer v_p.

As has been shown, the optimal flow rate in a nuclear resonance detector depends on the nonuniformity of the field measured, while the optimal flow rate in the polarizer and the connecting tube is constant. For this reason it is advantageous to pass part of the liquid from the connecting tube around the detector through a special bypass tube. Then any desired flow rate may be set in the detector, which is less than the flow rate in the polarizer. The maximum required flow rate in the detector is determined by the maximum possible nonuniformity of the measured field by means of expression (18.5):

$$q_{max} = v_a d_a \gamma (\operatorname{grad} H)_{max}. \tag{22.5}$$

The optimal liquid flow rate in the polarizer must be greater than or equal to this value

$$q_{p\,opt} \geqslant v_a d_a \gamma (\operatorname{grad} H)_{max}. \tag{23.5}$$

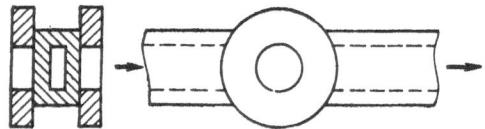

Fig. 3.5. Diagram of a slit flow detector for nuclear resonance.

By substituting $q_{p\ opt}$ from formula (19.5) in expression (23.5) we obtain

$$\frac{v_p}{T_1 \ln \frac{K+1}{K}} \geqslant v_a d_a \gamma (\text{grad } H)_{max} \qquad (24.5)$$

Hence, having set the value $K < 0.1$, it is possible to determine v_p. The value of K should be within the range 0.1-0.01. In this case the value of $\ln[(K + 1)/K] = 2.3 - 4.6$ and it may be taken as equal to 3.5 with an error of the order of 30%. The relaxation time T_1 for water equals 2 sec. Then

$$v_p \geqslant 7 v_a d_a \gamma (\text{grad } H)_{max}. \qquad (25.5)$$

Thus, we have determined in terms of the parameters of the detector and the field measured the polarizing volume, which is set by the volume of the interpolar space and, consequently, the dimensions and strength of the field of the polarizing magnet. The length of the connecting tube l_t must provide for a sufficient distance between the polarizing magnet and the field measured to avoid their interaction. It is set in the design in relation to the actual conditions. The diameter of the tube is determined from the condition $K \leq 0.1$ and thus

$$\frac{\pi d^2}{4} l_t \leqslant 0.1 v_p. \qquad (26.5)$$

The parameters of the nuclear resonance detector must be such as to guarantee the minimal measurement error, i.e., the minimal signal width and the maximal signal-to-noise ration. The signal width at the optimal flow rate is approximately equal to half the nonuniformity of the field measured in the volume of the detector and therefore, to decrease the width the detector should be as small as possible.

In a uniform external field the signal amplitude increases with an increase in the size of the detector. In a field with considerable nonuniformity it falls. As was shown previously, in practice a sufficiently good signal is observed in a cylindrical detector with an internal diameter of the tube of 3 mm and a coil length of about 4 mm, when the volume equals 0.03 cc.

For measuring nonuniform fields with a clearly marked gradient it is more convenient to use a slit detector, consisting of a channel with a rectangular section, to which are fixed two radiofrequency coils (Fig. 3.5). In such a detector a good signal is obtained with a channel cross section of 0.5 × 3 mm and an internal diameter of the coil of 3 mm. Here the width of the detector is 0.5 mm and its volume, 0.06 cc.

The detector should be arranged so that the gradient is along the axis of the coils (this is possible if the gradient is normal to the vector of the field strength). In this case the nonuniformity of the field in the detector is less by a factor of 6 than the nonuniformity of the field in the minimum possible cylindrical detector, while the signal amplitude in it is greater, probably because of the greater volume. This considerably increases the accuracy of the measurements.

The optimal flow rate is determined from expression (18.4) and the optimal amplitude of the oscillating field from formula (9.3).

4.5. Relation of the Error in the Measurement and Stabilization of a Field to Its Strength and Gradient

In a nonuniform external field the absolute error of the measurement and stabilization of the field is determined by expression (16.5). The relative error

$$\sigma_n = \frac{\text{grad } H}{H} \cdot \frac{d_a}{4a} . \qquad (27.5)$$

With an increase in the strength and gradient of the field, the signal amplitude and, consequently, the signal-to-noise ratio are proportional to H/(grad H) and therefore

$$\sigma_n \sim \left(\frac{\mathrm{grad}\, H}{H}\right)^2. \tag{28.5}$$

Thus, if the liquid flow rate q corresponds to q_{opt} for the minimal field gradient

$$q \approx \gamma\, \mathrm{grad}\, H_{min} d_a v_a,$$

then with an increase in the field, if its topography does not change, the relative error in the measurement and stabilization remains constant over the whole range of measurements, while if the topography changes, the relative error changes in proportion to the square of the relative nonuniformity of the field in the working volume of the detector. If the flow rate q and correspondingly the amplitude of the oscillating field are increased with increase in the nonuniformity of the field, remaining all the time optimal, then the signal-to-noise ratio a $\approx$ H and is independent of grad H and

$$\sigma_n \sim \frac{\mathrm{grad}\, H}{H^2}, \tag{29.5}$$

i.e., the error falls with an increase in the strength of the field measured.

As an example let us estimate the error in the measurement and stabilization of a field with a relative nonuniformity (grad H)/H = 0.01. With a detector with d_a = 0.3 cm, by using a phase detector it is possible to obtain a = 100 and this gives an error $\sigma_n \approx 10^{-5}$. In the same field with (grad H)/H = 0.1, the error will be 10^{-4}.

5.5. Application of Method

The method has been used for the measurement and stabilization of the magnetic field in a β-spectrometer of the "ketron" type in the range of 30-1000 oe [128]. The resonating substance was water from a town supply. An iron-clad electromagnet was used for magnetization. The diameter of the magnet was 450 mm and the height 600 mm, the diameter of the pole pieces was 200 mm, and the gap was 25 mm. With a current of 10 A the field in the gap was about 5000 oe and with a current of 25 A, about 10,000 oe. The strength of the stray field at a distance of 10 cm from the magnet was 4 oe and at a distance of 50 cm, about 0.1 oe. In the magnet gap was placed a chamber with a baffle and a spiral to prevent stagnation of the water and to increase the effective time for the water to pass through the polarizing field.

The detector was made in the form of a hollow cylinder of organic glass, on which was wound 100 turns of PÉ-0.2 wire in one layer. The length of the cylinder was 50 mm and the diameter 15 mm. These dimensions were determined by the size of the space in the "ketron" chamber intended for measuring the field. This choice of detector dimensions was based on the fact that with the nonuniformity which the measured field had at the position of the detector, with an increase in the detector volume the amplitude of the nuclear resonance signal increased more rapidly than its width, i.e., it was profitable to use a large detector to increase the accuracy of resonance tuning, while localization of the measurement region was not required.

The liquid flow rate was 50-100 cc/sec. The length of the connecting tube was 50 cm and the diameter 0.8 cm. The detector coil was connected to the oscillating circuit of a nuclear resonance detector of the autodyne type. With this apparatus it was possible to measure fields of 3 oe and above and to stabilize fields of 10 oe and above.

The error of the stabilization of the field was estimated from the constancy of the electron count on the steep side of an intense electron conversion line. In the region from 30 to 1000 oe it was no higher than $2 \cdot 10^{-5}$. The error in the measurement of the field was estimated from the reproducibility of the position of conversion lines in a β-spectrum. It was determined by the accuracy of the frequency measurement and at the layer level it was $3\text{-}4 \cdot 10^{-4}$.

CHAPTER 6

MEASUREMENT OF NONUNIFORM MAGNETIC FIELDS
(NUTATIONAL METHOD)

1.6. Comparison of Different Methods of Measuring Magnetic Fields

The simplest electromagnetic method [129-132], which is based on measurement of the emf induced by the magnetic field in a rotating coil, has an unlimited measurement range and small probe dimensions, but its readings are relative and the accuracy of the measurements does not exceed 0.1%.

The method of measuring field strength from the force acting on a conductor carrying a current placed in it is absolute and very accurate, but complicated. A high measurement accuracy (with an error of 10^{-3}%) may be obtained only in a uniform field with a long (up to 30 min) measurement process [133]. A simplified method makes it possible to carry out measurements with an error of 0.1-0.2% [134-137].

The Permalloy probe method [138-144] may be used for relative measurements of fields with a strength no higher than 100 oe. In fields less than 1 oe it gives an error of 10^{-5} and in fields of 10-100 oe the measurement error is 10^{-4} because of the nonuniformity of the compensating field. The method is relative and the calibration of the instruments is upset by the smallest change in the geometry of the detectors and by the effect of ferromagnetic masses nearby. The projection of the field along the axis of the compensating coil is measured and therefore a change in the topography of the field necessitates reorientation of the detector.

Semiconductor measuring instruments for fields based on the Hall effect [145-151] have a very low accuracy (2%). Magnetometers with a bismuth spiral [152] may have an accuracy down to 0.1%, but they cannot be used in fields below 1000 oe. Magnetometers based on the magnetoconcentration effect [153] have a wide range of application, but do not give measurements of high accuracy. The common drawback of semiconductor measuring instruments is the dependence of the readings on temperature and other parameters so that they can only be used for rough measurements.

The paramagnetic resonance method has a low accuracy as, firstly, the g-factor of electrons in a substance depends on the temperature and other parameters and, secondly, an electron resonance line has considerable width. For example, the organic free radical α,α-diphenyl-β-picrylhydrazine (DPPH) has a line width of 1.7 oe, while α-(p-chlorovinyl)-α-phenyl-β-picrylhydrazine (CPPH) has a line width of 2 oe [154], and other organic free radicals have even wider lines. With some inorganic free radicals the lines are narrower. For example, a solution of sodium in liquid ammonia has a line with a width of 0.1 oe [155], but this gives a signal amplitude which is less by a factor of 60 than with the same amount of DPPH. There are field measuring instruments using both DPPH [154-156] and a solution of sodium in ammonia [154, 156] and they give a measurement accuracy of the order of 10^{-3}.

The nuclear magnetic resonance method with a flow detector and preliminary polarization makes it possible to measure a field in the range of a few oersteds and above with a detector volume of 0.01-0.03 cc. The permissible relative nonuniformity of the field with strengths above 300 oe is 1-2% per cm and in weak fields, 20% per cm. The advantages of the method are the fact that it is absolute and has a high accuracy, while the drawbacks are the low permissible nonuniformity of the field, which is limited by the lower limit of the strength of the measureable fields, and the need for changing detectors for measurements over a wide range.

For using the nuclear magnetic resonance method in less uniform fields it is possible to compensate artificially for the gradient of the external field by a system of conductors carrying a current [157]. This makes it possible to measure the strength of a field with a gradient of 1000-1200 oe/cm with an accuracy of 0.01%.

The gradient compensation method does not solve the problem of the absolute measurement of nonuniform fields as to use it, it is necessary to observe the nuclear resonance signal in the nonuniform field and then, by selecting the currents in the system of conductors, to compensate for the gradient, increasing the signal amplitude. Therefore, the maximum permissible gradient is determined not by the gradient of the field, measured with compensation, but by the gradient of the field in which it is possible to detect a nuclear resonance signal without compensation.

At the present time two more methods of measuring the strength of a magnetic field have been proposed. A hydrogen ion meter [158, 159], which uses the cyclotron resonance of ionized hydrogen molecules H_1^+, H_2^+ and H_2^+, H_3^+, has a measurement range of 1000-31,500 oe with a nonuniformity up to 60 oe/cm. The accuracy of the measurements is $5 \cdot 10^{-4}$, the detector volume 2.2 cc, and the dimensions of the detector are $20 \times 36 \times 25$ mm. The drawback of the method is the large size of the detector.

An electron resonance meter [160, 161], which uses the cyclotron resonance of electrons, has a measurement range of 1-12 oe and a relative accuracy of $5 \cdot 10^{-4}$. The working volume of the detector is 19 cc and the dimensions of the detector are $40 \times 50 \times 150$ mm. The method is relative and therefore it is inferior to the Permalloy method in which the volume of the detector is much less and the sensitivity higher. A similar method was proposed in the work of other authors [162].

From the review presented it may be concluded that the most universal method, which is suitable for measuring magnetic fields over a wide range of strengths and with high gradients without changing the detector, is the electromagnetic method, which, however, is relative and has a low accuracy. All the other methods have a limited range of application and to increase the range it is necessary to change the detectors. There are two absolute methods among these: the nuclear resonance method, which may be used for measuring fields of 3 oe and above with 4-5 changeable detectors, and the hydrogen cyclotron resonance method, which may be used in fields with a strength above 1000 oe with a detector volume of about 2 cc. Neither of these methods are applicable in fields with high nonuniformity.

As already mentioned, under certain conditions the amplitude of a nutation signal is independent of the strength of the magnetic field in the nutation detector and the dimensions of the detector. This makes it possible to obtain a nutation signal in fields of any strength and with a detector of small volume. In practice, a nutation signal has been observed in fields with a strength of 0.007-20,000 oe with a nutation detector volume of 0.03 cc and above. Moreover, if the direction of the liquid flow in the nutation detector coincides with the direction of the gradient of the external field, then with a sufficiently high flow rate and strength of the oscillating field, the amplitude of the nutation signal is independent of the nonuniformity of the external field. If in the nutation detector it is impossible to eliminate the nonuniformity of the field at right angles to the liquid flow, then its effect may be reduced considerably by using modulation of the external field. This makes it possible to obtain a nutation signal in strongly nonuniform fields (with a relative nonuniformity more than 100% per cm).

Thus, the method makes it possible to make absolute measurements of magnetic fields over a wide range of strengths and with high gradients using standard nuclear resonance apparatus without changing the detector. The error associated with the shift of the nuclear resonance signal with considerable changes in the velocity of the liquid [195] may be eliminated by judicious design of the detector.

2.6. Estimation of Accuracy of Measurement of Magnetic Field

The accuracy of the measurement of a magnetic field depends on the accuracy of the determination of the frequency at which the maximum nutation effect is observed. A nutation signal has quite a flat apex, whose center is difficult to fix directly and therefore, in practice the highest accuracy is achieved by determining the points at which the absorption signal changes polarity. If these points are fixed with an error of $\pm \Delta H$, then the central point between them may be determined with the same error. This value is the absolute error of the measurement of the field. Let us denote the amplitude of the noise in the detector circuit by A_{no}. The appearance of an absorption signal of either polarity may be detected only when A is comparable with A_{no}. A typical form of a nutation signal is shown in Fig. 1.6. With an error of no more than 50% it may be assumed that at the points of inversion of the absorption signal the derivative $dA/dH = 2A_{max}/\delta H$, where A_{max} is the

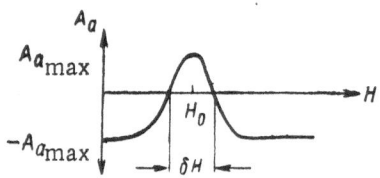

Fig. 1.6. Form of nutation signal.

maximum amplitude of the absorption signal and δH is the width of the nutation signal at the half-height. At the points of inversion of the absorption signal $A = 0$ and with a shift in the field of $\pm \Delta H$ there appears an absorption signal with an amplitude $A = (dA/dH) \Delta H$. The value ΔH at which $A = A_{no}$ determines the measurement error and it equals

$$\Delta H = \frac{A_{no}}{dA/dH} = \frac{\delta H}{2a} , \tag{1.6}$$

where $a = A_{max}/A$ is the signal-to-noise ratio.

Thus, to obtain the minimal measurement error it is necessary to ensure that the nutation signal has the minimum width. With a nutation detector of large dimensions, δH is determined by the nonuniformity of the external field and therefore it falls with an increase in the volume of the detector. With small dimensions, the width of the signal is determined by the strength of the oscillating field and the instrument effect and falls with an increase in the volume of the detector. Thus, with a given nonuniformity of the field there is some optimal volume of the nutation detector at which the width of the signal is minimal.

As was shown in Section 2.2, in a nonuniform field with l_n grad $H \gg q/\gamma v_n$

$$\delta H = \frac{l_n \operatorname{grad} H}{2} ,$$

where l_n is the length of the nutation detector. By substituting this value in expression (1.6) we obtain the relation of the absolute measurement error to the nonuniformity of the field

$$\Delta H = \frac{l_n \operatorname{grad} H}{4a} . \tag{2.6}$$

The relative error

$$\sigma_n = \frac{\Delta H}{H} = \frac{l_n}{4a} \cdot \frac{\operatorname{grad} H}{H} . \tag{3.6}$$

When $l_n = 4$ mm and $a = 100$

$$\sigma_n = 10^{-3} \frac{\operatorname{grad} H}{H} .$$

If the relative nonuniformity of the field $(\operatorname{grad} H)/H = 10^{-2}$, then $\sigma_n = 10^{-5}$ (with this nonuniformity of the field, a Permalloy probe gives an error of 10^{-4}) [144].

The measurement error may be reduced considerably by artificial compensation for the gradient of the external field [157]. In this case, the nutation method, in contrast to the normal nuclear resonance method, may be used in fields with a very high nonuniformity as a nutation signal may be observed with a much higher gradient of the external field in the detector than an absorption signal.

As was shown in Section 2.2, in a uniform field (when l_n grad $H \ll q/\gamma v_n$) the signal width

$$\delta H = \frac{1.8 \pi q}{\gamma v_n} ,$$

whence the absolute error of the measurements $\Delta H = \pm\, 0.9\pi q/\gamma v_n a$. When $v_n = 0.2$ cc, q = 20 cc/sec, and a = 100, $\Delta H = 10^{-4}$ oe. This relation of the signal width to the parameters of the detector was observed with a cylindrical flow detector. With a change in the form of the detector it is possible to change only the broadening due to movement of the nuclei, but the contribution of this broadening to the width of the nutation signal is only 14%, i.e., even with a change in the broadening by 50% the width of the signal changes by only 7% so that the expression for the measurement error may be regarded as valid for any form of flow detector.

In measuring the strength of weak magnetic fields with a detector of small volume it is necessary to take into account the correction for the Bloch–Siegert shift [95-97, 163], which is given by

$$\Delta_B = -\frac{H_1^2}{4H} = -\frac{\pi^2 q^2}{4\gamma^2 v_n^2 H} \, .$$

When H = 10 oe, $v_n = 0.3$ cc, and q = 30 cc/sec, this correction is small and $\Delta_B = -4 \cdot 10^{-6}$ oe, but when H = 0.1 oe, $v_n = 0.03$ cc, and q = 30 cc/sec, $\Delta_B = -0.04$ oe. With this correction taken into account, the relation of the field strength to the resonance frequency of the oscillating field has the form

$$\gamma H = \frac{\omega}{2}\left(1 + \sqrt{1 - \frac{\pi^2 q^2}{v_n^2 \omega^2}}\right) = \omega\left(1 - \frac{\pi^2 q^2}{4 v_n^2 \omega^2}\right). \tag{4.6}$$

3.6. Optimal Dimensions of Nutation Detector

As has been shown, with a nutation detector of large dimensions, when $l_n \operatorname{grad} H \gg \pi q/\gamma v_n$, the absolute error of the measurement of a magnetic field

$$\Delta H = \frac{l_n \operatorname{grad} H}{4a} \, .$$

With a detector of small size, when $l_n \operatorname{grad} H < q/\gamma v_n$, the absolute error of the measurement $\Delta H = 0.9\, \pi g/v_n\, a$.

Thus, with a detector of large size the measurement error increases with an increase in the volume of the detector, while with the small detector, the error increases with a decrease in the volume of the detector. Consequently, there are some optimal dimensions of the detector with which the measurement error is minimal.

It is obvious that the optimal dimensions must satisfy the condition

$$l_{n\,\text{opt}}\,\frac{\operatorname{grad} H}{4a} = \frac{0.9\,\pi q}{\gamma v_{n\,\text{opt}} a} \, . \tag{5.6}$$

Hence, by substituting $v_n = (\pi d_{n\,\text{opt}}^2/4)l_{\text{opt}}$, where d_n is the diameter of the detector, we obtain

$$l_{n\,\text{opt}}^2\, d_{n\,\text{opt}}^2 = \frac{15q}{\gamma \operatorname{grad} H} \, . \tag{6.6}$$

If $l_n = d_n$, then

$$d_n = \sqrt[4]{\frac{15q}{\gamma \operatorname{grad} H}} \, . \tag{7.6}$$

With a detector of rectangular form with the dimensions l_n, c_n, and d_n, the optimal dimensions are related by the expression

$$l_{n\,opt}^2 c_{n\,opt} d_{n\,opt} = \frac{12q}{\gamma\,\text{grad}\,H}.\tag{8.6}$$

In all cases l is the length of the detector along the gradient of the field.

4.6. Practical Designs of Nutation Detectors

Let us examine some designs of nutation detectors. The simplest design consists of a glass or plastic tube with a cylindrical coil on it. A detector with the diameter of the working volume equal to the diameter of the connecting tube (Fig. 2a.6, see Fig. 3a.3) is convenient to use in fields with a high gradient and a detector with the diameter of the working volume greater than the diameter of the connecting tube (see Fig. 2b.6) should be used in uniform fields. This design has two drawbacks: 1) it is suitable only for fields with a gradient at right angles to the field as in the measurement it is necessary for the liquid to flow parallel to the gradient, while the direction of the axis of the coil should be perpendicular to the vector of the strength of the external field; 2) with this design it is impossible to make the length of the detector less than its diameter as the field of the coil extends along its axis a distance greater than its diameter.

Designs of detector in the form of a tube with a transverse flat coil are shown diagramatically in Fig. 3.6. The coil is wound on a flat form, whose thickness determines the length of the field along the axis of the tube.

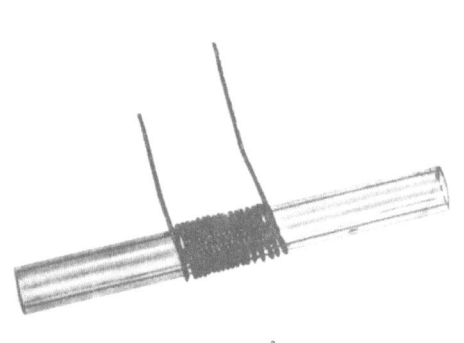

The coil is enclosed in a brass screen to limit the spread of the oscillating field along the axis of the tube. If the diameter of the detector equals the diameter of the connecting tube (see Fig. 3a.6), the detector is suitable for a high field gradient. When the diameter of the detector is greater than the diameter of the connecting tube (see Fig. 3b.6), the detector is suitable for use in uniform fields. If the working volume has a conical form (see Fig. 3c.6), by moving the coil along the tube it is possible to change the transverse dimensions of the detector, adjusting them to the optimal values for the nonuniformity of the field. The section of the detector tubes (see Fig. 3.6) may be circular or rectangular. As the axes of the tube and the coil are perpendicular, it is possible to use

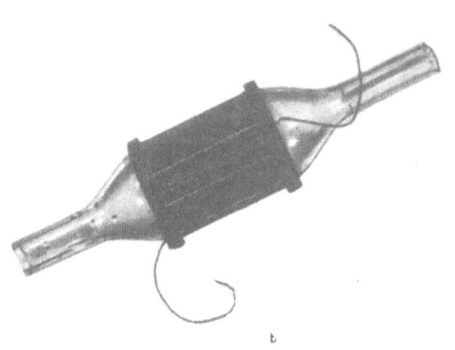

Fig. 2.6. Nuclear resonance detectors; a) for nonuniform magnetic fields; b) for uniform fields.

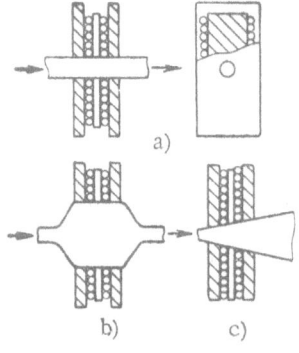

Fig. 3.6. Designs of nutation detectors with a transverse flat coil.

the detector with any relative direction of the vector and the gradient of the external field. The length of the working volume equals the thickness of the coil (the distance between the brass screens).

5.6. Application of Nutation Method

Tuning of Nuclear Resonance Detectors. In constructing nuclear resonance detectors it is very difficult to obtain a signal first time as it is necessary to have sufficiently high uniformity of the external field in the detector coil, to set the frequency of the oscillator of the detector at the nuclear precession frequency, and to establish the optimal amplitude of the modulating field. As there are no methods of control, the adjustment to optimal conditions for resonance in the detector is normally done blindly and this leads to great difficulties, especially if experience is lacking. The use of the nutation effect considerably eases this problem.

The detector of a tuned NMR detection system is included in the flow system between the polarizers and the other flow detector in which the nuclear resonance signal is observed. The potential from the oscillator is fed to the coil of the first detector and the nuclear precession frequency determined from the change in the signal in the second detector. The uniformity of the field at the position of the detector is estimated from the width of the nutation signal and if it is insufficient to obtain a signal, the topography of the magnetic field is smoothed out and a point with sufficient uniformity found, using the nutation signal for monitoring. The detector is fixed at this point and the oscillator frequency set at the nuclear precession frequency (from the maximum of the amplitude of the nutation signal) and then the optimal amplitude of the potential on the detector coil selected at which the nutation angle equals $3\pi/4$, i.e., the amplitude of the signal in the second detector equals 0.7 of the amplitude in the absence of nutation in the first coil. It can now be assumed with certainty that in the detector of the tuned NMR detection system there are conditions for obtaining the maximum signal and regulation of the detector circuit begun.

Experiments on the Use of Nuclear Resonance in Magnetic Flaw Detection. One of the promising methods of magnetic flaw detection is the magnetic probe method with small elements for measuring the magnetic field strength.

There are several types of probe with different qualities and characteristics. Here we present the results of experiments on the use of a nuclear magnetic resonance detector as the sensitive element.

The main advantage of measuring a magnetic field by nuclear magnetic resonance lies in the fact that with any orientation of the detector the mean value of the vector of the field strength through its volume is determined. All other forms of probe give readings depending on the orientation of the sensitive element relative to the direction of the vector of the magnetic field and this considerably hampers the measurements, especially at surfaces of varying curvature.

Another, no less important advantage is the possibility of measuring the field very close to the surface of an object (less than 1 mm away) because of the small size of the nuclear resonance detector and the fact that the measurements are absolute. In the most refined Permalloy gauges, to obtain a high sensitivity it is necessary to use a compensating coil of large dimensions, which determines the size of the sensitive element. Moreover, bringing this coil near to the surface of the object upsets the calibration of the gauge. Therefore, in practice it is impossible to measure the field close to the surface of an object with high accuracy with Permalloy detectors. At the same time, it is obvious that the distortion of the stray magnetic field due to the presence of a flaw falls sharply with an increase in the distance from the surface.

The magnetic field close to the samples investigated is weak and very nonuniform and therefore the nutation method was used to measure it. The sensitive element of the probe was the first detector (nutation detector), 0.2—0.4 cm in size. In this apparatus H_p

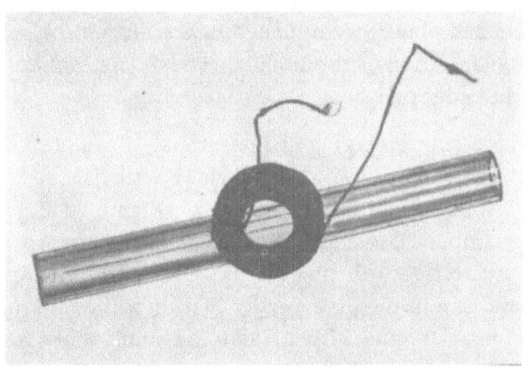

Fig. 4.6. Nutation detector.

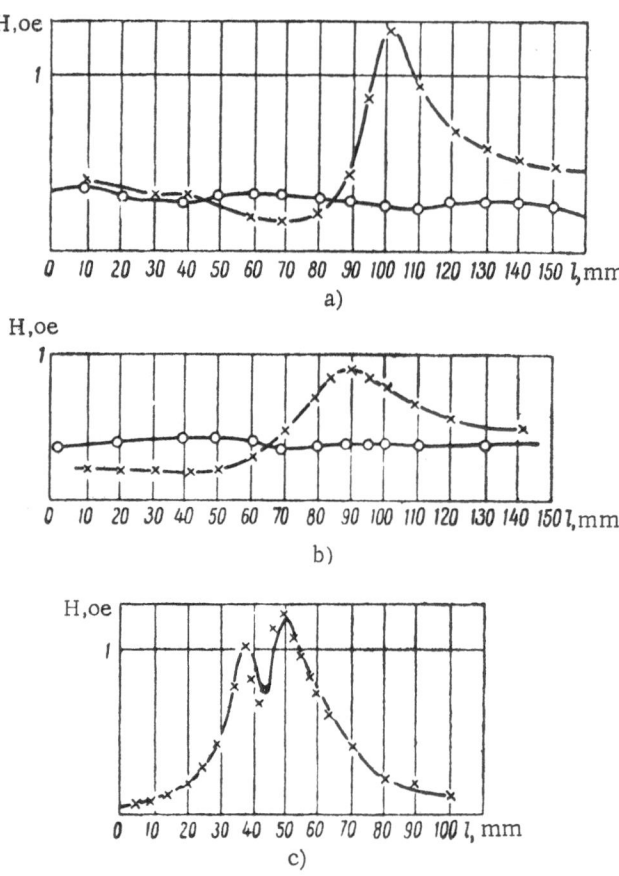

Fig. 5.6. Topography of the magnetic field at the flat surface of steel objects with artificial flaws; a) with one simulated crack; b) with one simulated crack that did not emerge at the surface; c) with two simulated cracks.

= 10,000 oe, v_p = 400 cc, and H_a = 30 oe. The length of the connecting tube was 200 cm and its cross section 0.07 cm^2. The nutation detector is shown in Fig. 4.6. By means of it, it was possible to measure magnetic fields very close to the surface of objects. The volume of the nutation detector was 0.03 cc and that of the absorption detector 5 cc. The absorption signal was detected with an autodyne circuit and the field was measured with an error of 0.004 oe.

For carrying out the experiment the nutation detector was fixed in a special clamp and moved along the surface of the samples examined in the chosen direction. The data from some measurements are given in Fig. 5.6. The samples examined, which were polished steel bars and plates, were placed in the magnetic circuit of a small magnet (the same magnet could be used for polarization of the liquid). The strength of the magnetic field at a distance of 5 mm from the sample was of the order of 1 oe.

For determining the sensitivity of the method for revealing cracks emerging at the surface, a sample was prepared which consisted of a polished steel bar with a section of 30 × 30 mm, onto which were placed plates 9 mm thick with end faces whose surfaces were finished with an accuracy of about 1 μ. Figure 5a.6 gives the measurement data obtained by moving the probe along this sample. The high peak on the curve corresponds to the simulated crack, which consisted of the boundary between two tightly fitting plates. The curve without the peak corresponds to one plate (without cracks) and is the base line in the given case. Data from measuring the strength of the magnetic field when the probe was moved along a sample with a simulated crack which did not emerge at the surface are given in Fig. 5b.6. The peak corresponded to the junction of two plates, covered with a 3-millimeter steel plate. The figure also gives the base line (without a flaw). The curve in Fig. 5c.6

corresponds to a sample with two artificial cracks lying close to each other. The figure shows that with the given distance between the cracks, the flaws are resolved.

The experiments confirm the possibility of using a nuclear field meter as the sensitive element of a magnetic probe for revealing defects in objects made of ferromagnetic materials, including pinched cracks which do not emerge at the surface.

Stabilization of Magnetic Fields. For the stabilization of a magnetic field by the nutation method, in this field is placed a nutation detector fed by an oscillator of high stability. The amplitude of the resonance oscillating field is set at the optimal value for obtaining the first nutation extremum and the frequency of the oscillator shifted relative to exact resonance so that the amplitude of the absorption signal equals zero. With a change in the strength of the field stabilized, the nutation angle changes and at the output of the detector circuit connected to the absorption detector there appears a signal, whose polarity depends on the direction of the change in the field stabilized. By feeding this signal to a stabilization circuit it is possible to compensate for the change in the field stabilized. An advantage of the method is the fact that the nutation detector does not contain a resonance circuit and therefore it can be used over an unlimited range of frequencies without retuning. The amplitude of the nutation signal is independent of the field strength and depends little on its nonuniformity so that the method can be used in weak and nonuniform fields.

A quartz oscillator may be used as a frequency source, while a smooth change in the field may be achieved with an auxilliary audiofrequency oscillator, operating on the nuclear resonance side band. The method has been used for stabilization of the magnet field of a $\pi\sqrt{2}\,\beta$-spectrometer with a nonuniformity of 1% per cm.

CHAPTER 7

MEASUREMENT OF LIQUID FLOW RATES THROUGH THE NUCLEAR RESONANCE SIGNAL AMPLITUDE

1.7. Essence and Characteristics of Measurement Method

In connection with the development of new fields of science and technology there has arisen a need for methods for the remote measurement of flow rates of aggresive liquids. To some extent these requirements are satisfied by an electromagnetic flowmeter [167, 168] in which the liquid flow rate is measured through the emf produced in a conducting liquid when it flows through a magnetic field. This method is used widely in practice. An electromagnetic flow meter may be used only with conducting liquids and up to now, no such instruments for measuring the flow of liquid dielectrics have existed. Nuclear magnetic resonance makes it possible to solve this problem.

A schematic diagram of the apparatus is shown in Fig. 1.7 [37]. A section of the tube with a volume v_p is placed in a strong polarizing field H_p. The end of this section with a volume v_a, lying in a field H_a, is the nuclear resonance detector, the coil of the circuit of which is connected to a detection system. The liquid is polarized in the field H_p and in flowing through the detector it gives a nuclear resonance signal, whose amplitude A depends on the liquid flow rate. Naturally, such an apparatus may be used for measuring the flow rate of liquids with a large number of nuclei giving a good nuclear resonance signal. The concentration of the nuclei must remain constant with an accuracy greater than the specified error in the measurement of the flow rate or the change in it should be known for introducing a correction into the sensitivity of the instrument. The concentration of the nuclei is proportional to the density, which is a function of temperature, i e., the sensitivity is related to the temperature. The effect of a change in the other parameters of the liquid may be eliminated by the design of the instrument.

The working section of the tube must be made of a nonmagnetic material and the section inside the detector coil must not be electrically conducting. In practice it is most convenient to use plastic.

Information on the flow rate is obtained in the form of an electric potential of low frequency, which may be transmitted for a distance and also used in control and regulation systems. The flowmeter has no moving parts. The system may be hermetically sealed and this is particularly convenient with poisonous and active liquids and the flowmeter does not produce any additional pressure losses. It is insensitive to a change in the orientation of the tube in space. It has a linear scale. Its time lag is determined by the time constant of the recording system and may be made quite low. Without much complication in the design, the instrument may be calibrated periodically by the method for the absolute measurement of liquid flow rates described in Ch 8.

2.7. Optimal Parameters of Instrument and Measurement Error

The parameters of the flowmeter must be selected so that the nuclear resonance signal has the maximum amplitude, the readings are independent of the relaxation time of the liquid T_1, and the scale of the instrument is linear. For this purpose it is first necessary that the magnetization of the nuclei in the liquid entering the detector should be maximal and independent of the relaxation time T_1. If the strengths of the fields in the polarizer and detector are equal, $H_p = H_a$, the magnetization is determined by the expression

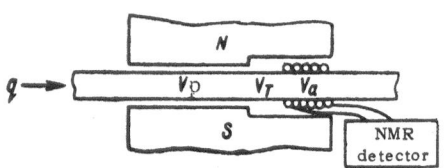

Fig. 1.7. Diagram of amplitude nuclear resonance flowmeter.

$$M = X_0 H_p \left(1 - e^{-\frac{v_p}{q T_1}}\right).$$

If $H_a \neq H_p$, then to ensure that the field H_a is uniform the NMR detector should lie at some distance from the edges of the fields H_p and H_a. Let us denote by v_t the volume of the section of the tube through which the liquid flows from the polarizer into the NMR detector, lying in the intermediate field H_t. In this case the magnetization is determined by the expression

$$M = X_0 H_p \left(1 - e^{-\frac{v_p}{qT_1}}\right) e^{-\frac{v_t}{qT_1}} + X_0 H_t \left(1 - e^{-\frac{v_t}{qT_1}}\right) \tag{1.7}$$

or

$$M = X_0 H_p (1 + B),$$

where

$$B = \left(\frac{H_t}{H_p} - 1\right)\left(1 - e^{-\frac{v_t}{qT_1}}\right) - e^{-\frac{v_p + v_t}{qT_1}} . \tag{2.7}$$

The value M appears as a cofactor in the expression for the amplitude of the NMR signal and to guarantee measurement of the flow rate with a relative error less than $\sigma_{q\,per}$, over the whole range of permissible changes in the relaxation time T_1 the relative change in M should no exceed $\sigma_{q\,per}$. This may be guaranteed if the following condition is fulfilled:

$$B < \sigma_{q_{per}} . \tag{3.7}$$

With a flow rate q_{min} at the lower limit of the range of measurements the main contribution is made by the first term and condition (3.7) will be

$$\left[\left(\frac{H_t}{H_p} - 1\right)\left(1 - e^{\frac{v_t}{q_{min}T_1}}\right)\right] < \sigma_{q_{per}} .$$

Thus we obtain the condition for the selection of v_t

$$v_t < q_{min} T_1 \ln\left(1 - \sigma_{q\,per} \frac{H_p}{H_p - H_t}\right) . \tag{4.7}$$

With a flow rate q_{max} at the upper limit of the range of measurements the main contribution to B is made by the second term and expression (3.7) assumes the form

$$e^{-\frac{v_p + v_t}{qT_1}} < \sigma_{q\,per} .$$

Hence we obtain the condition for selecting the volume of the polariser

$$v_p + v_t > q_{max} T_1 \ln \frac{1}{\sigma_{q_{per}}} . \tag{5.7}$$

The amplitude of the signal in the detector is determined by expression (5.3), which includes the factor $\gamma H_1 Z T_{1n} T_{2n}$ that depends on the relaxation time T_{1n}. The latter may change with a change in temperature, the viscosity and composition of the liquid, and also its flow conditions. To eliminate the dependence on T_1 of this factor, it is essential that

$$\gamma^2 H_1^2 T_{1n} T_{2n} \gg 1,$$

then

$$Z = \frac{1}{\gamma^2 H_1^2 T_{1n} T_{2n}} \left(1 - \frac{1}{\gamma^2 H_1^2 T_{1n} T_{2n}} \right)$$

and the factor mentioned above becomes equal to

$$\frac{1}{\gamma H_1} \left(1 - \frac{1}{\gamma^2 H_1^2 T_{1n} T_{2n}} \right) .$$

For the error introduced by this factor with a change in T_{1n} not to distort the results of the measurements, it is necessary to have the condition

$$\frac{1}{\gamma^2 H_1^2 T_{1n} T_{2n}} < \sigma_{q\,per}$$

or

$$Z < \frac{\sigma_{q\,per}}{1 + \sigma_{q\,per}} . \qquad (6.7)$$

When $Z \ll 1$, $\gamma H_1 T_{2n} \ll 1$ and $T_{2n} \ll T_{1n}$, expression (5.3) has the form

$$A = \frac{A_M q}{\gamma H_1 v_a} \left(1 - e^{-\frac{v_a}{q T_{1n} Z}} \right) + \frac{A_M H_a}{\gamma H_1 T_1 H_p} . \qquad (7.7)$$

As the condition $\gamma H_1 T_{2n} \ll 1$ is valid with a nonuniformity of the field in the detector greater than H_1, in expression (7.7) one should take into account the decrease in the effective filling coefficient of the detector with resonating nuclei (Section 1.3), but this is not important for subsequent discussions.

The following two conditions should be fulfilled for the relation of the signal amplitude to the flow rate to be unaffected by the relaxation time:

$$q_{max} < \frac{v_a}{T_{1n} Z \ln \frac{1}{\sigma_{q\,per}}} , \qquad (8.7)$$

$$q_{min} < \frac{v_a H_a}{T_1 H_p \sigma_{q\,per}} . \qquad (9.7)$$

When $Z \ll 1$, $\gamma H_1 T_{2n} \gg 1$ and $T_{2n} \ll T_{1n}$, expression (5.3) has the form

$$A = \frac{A_M q}{\gamma H_1 v_a} \left[1 - \left(\cos \gamma H_1 \frac{v_a}{q} + \frac{\sin \gamma H_1 \frac{v_a}{q}}{\gamma H_1 T_{2n}} \right) e^{-\frac{v_a}{q T_{2n}}} \right] + \frac{A_M H_0}{\gamma H_1 H_p T_1} \left(1 - \frac{\sin \gamma H_1 \frac{v_a}{q}}{\gamma H_1 \frac{v_a}{q}} e^{-\frac{v_a}{q T_{2n}}} \right) . \qquad (10.7)$$

For the signal amplitude to be independent of the relaxation time, the following conditions should be fulfilled:

$$q > \frac{v_a}{T_{2n} \sigma_{q\,per}} \tag{11.7}$$

or

$$q < \frac{v_a}{T_{2n} \ln \frac{1}{\sigma_{q\,per}}}, \tag{12.7}$$

and also condition (9.7). In practice, of the conditions (11.7) and (12.7), only the latter is realistic as condition (11.7) is upset by the instrument effect. The value of T_{2n} with transverse nonuniformity of the field in the detector $\Delta H_\perp$ approximately equals $T_{2n} = 2/\gamma \Delta H_\perp$. By substituting this value in condition (12.7) we obtain when

$$\frac{H_1}{\Delta H_\perp} \gg 1,$$

$$q_{max} < \frac{v_a \gamma \Delta H_\perp}{2 \ln \frac{1}{\sigma_{q\,per}}}. \tag{13.7}$$

By substituting the same value in condition (8.7), we obtain

$$\frac{H_1}{\Delta H_\perp} \ll 1$$

$$q_{max} < \frac{2 v_a \gamma H_1^2}{\ln \frac{1}{\sigma_{q\,per}} \Delta H_\perp}. \tag{14.7}$$

For conditions (13.7) and (14.7) it follows that the range of the measurements may be extended by increasing $\Delta H_\perp$ when $H/\Delta H_\perp \gg 1$ and by decreasing $\Delta H_\perp$ when $H_1/\Delta H_\perp \ll 1$ or by increasing H_1. The most favorable condition is when $H_1 \approx \Delta H_1$ or $2\gamma H_1 T_{2n} \simeq 1$. When $Z \ll 1$, $2\gamma H_1 T_{2n} = 1$ and $T_{2n} \ll T_{1n}$, expression (5.3) has the form

$$A = \frac{A_M q}{\gamma H_1 v_a} \left[1 - \left(1 + \frac{v_a}{q T_{2n}} \right) e^{-\frac{v_a}{2q T_{2n}}} \right] + \frac{A_M}{\gamma H_1 T_1} \cdot \frac{H_0}{H_p} \left(1 - e^{-\frac{v_a}{2q T_{2n}}} \right). \tag{15.7}$$

In this case, one of the conditions for the readings to be independent of the relaxation time will be (9.7) and the other

$$\left(1 + \frac{v_a}{2q T_{2n}} \right) e^{-\frac{v_a}{2q T_{2n}}} < \sigma_{q\,per}. \tag{16.7}$$

For example, when $\sigma_{q\,per} \approx 0.01$, from the inequality (16.7) it follows that

$$q_{max} \leqslant 0.04 \, v_a \gamma \Delta H_\perp,$$

and when $\sigma_{q\,per} = 0.05$,

$$q_{max} \leqslant 0.05 \, v_a \gamma \Delta H_\perp.$$

By substituting in the inequality (6.7) $\gamma H_1 \approx \gamma \Delta H_\perp$ and $T_{2n} \approx 2/\gamma \Delta H_\perp$, we obtain the condition for estimating the required nonuniformity of the field in the detector

$$\gamma \Delta H_\perp T_{1n} \geqslant \frac{1}{2\sigma_{q\,per}} . \tag{17.7}$$

Let us also consider the selection of the strength of the field H_a in the NMR detector. When $H_a = H_p$, expression (9.7) sets the lower limit of the range of measurements, while when $H_a \ll H_p$ it is necessary to remove the detector away from the polarizing field and then there is the lower limit of the range of measurements set by expression (4.7). In practice, to guarantee sufficient uniformity of the field H_a in the region of the detector the length of the connecting section of the tube, lying in a field $H_t \approx H_a$, must be not less than 10 diameters of the tube, i.e., $v_t \approx 10v_a$. In the general case

$$v_t \approx k v_a,$$

where k depends on the actual construction of the magnetic system. At the lower limit of the range of measurements, using expressions (4.7) and (9.7), the following conditions must hold:

$$v_a \leqslant q_{min} T_1 \frac{H_p}{H_a} \sigma_{q\,per}$$

and

$$v_t \leqslant q_{min} T_1 \sigma_{q\,per} \frac{H_p}{H_p - H_a} .$$

The volume v_a should be as large as possible to increase the signal amplitude and the volume v_t should be as large as possible so that the detector is at a distance from the polarizing field. Assuming that $v_t = k v_a$, we obtain

$$H_a = \frac{H_p k}{k+1} . \tag{18.7}$$

Physically this relation means the following. The demagnetization of the liquid in the volume v_t is described by the factor $\left[1 + \frac{H_a - H_p}{H_p} \left(1 - e^{-\frac{v_t}{q T_1}} \right) \right]$, and the parasitic magnetization of the liquid in the volume v_a is described by the factor $(1 + (H_a/H_p)(v_a/T_2))$, so that the product of these factors, allowing for the smallness of $v_t/q T_1$, equals

$$1 - \frac{v_t}{q T_1} \left(1 - \frac{H_a}{H_p} \right) + \frac{H_a}{H_p} \cdot \frac{v_a}{q T_1} .$$

Terms of the second order of smallness are eliminated. This expression equals unity when we have the condition

$$v_t \frac{H_p - H_a}{H_p} = \frac{H_a}{H_p} v_a.$$

By substituting $v_t = k v_a$ we obtain

$$H_a = \frac{H_p k}{k+1} .$$

96

Thus, by an appropriate choice of the strengths of the fields H_p and H_a it is possible to eliminate the contribution of the first order of smallness in the measurement error and then the product will equal

$$1 + \frac{H_a}{H_p} \cdot \frac{(k-1)_t v_a^2}{2q^2 T_1^2}$$

(terms of the third order of smallness are eliminated).

Thus, instead of conditions (4.7) and (9.7) we have a new condition which holds provided that condition (18.7) holds:

$$\frac{k(k-1)v_a^2}{2(k+1)^2 q^2 T_1^2} \leq \sigma_{q\,per},$$

i.e.,

$$q_{min} \geq \frac{v_a k}{T_1 (k+1)} \sqrt{\frac{k(k-1)}{2\sigma_{q\,per}(k+1)}}$$

(19.7)

3.7. Example of Practical Calculation

We have set the limits of the flow rates q_{min} = 1 cc/sec and q_{max} = 10 cc/sec, the measurement error 0.01, and the relaxation time $T_1 \approx 1$ sec. If 0.01 is the error relative to the upper limit of the scale, then with q = 1 cc/sec the actual error $\sigma_{q\,per}$ = 0.1 and by assuming that k = 10 and using condition (19.7), we obtain the volume of the detector v_a = 0.15 cc and $d_a = \sqrt[3]{v_a}$ = 5.3 mm. The volume v_t = 1.5 cc and the field $H_a = H_t$ = $H_p 10/11$, so that with H_p = 5000 oe, H_a = 4550 oe. The nonuniformity of the field from expression (16.7) must be greater than 0.12 oe/cm and from the inequality (17.7) with $T_{1n} \simeq 0.01$ sec it must be greater than 0.35 oe/cm.

From expression (5.7) we find the volume of the polarizer

$$v_p = 45 \text{ cc.}$$

As the accuracy of the flow rate measurement cannot exceed the signal-to-noise ratio, with the minimum flow rate q_{min} this ratio should not be less than 10. Resonance conditions should be maintained strictly during the measurements and the relative change in the amplitude due to a shift of the frequency or the field should not exceed $\sigma_{q\,per}$. With the same accuracy it is necessary to maintain the amplitude of the oscillations in the circuit for H_1, the stability of the polarizing field, and the potential in the receiver circuit. As the magnetic susceptibility of the nuclei X_0 depends on the temperature and the chemical composition of the substance, the change in these parameters must be monitored and appropriate corrections made to the readings of the instrument.

4.7. Application of Amplitude Method for Measuring Blood Flow Rate

A study of the principles of blood flow in vessels of a living organism can give very much valuable information on the function of the cardiovascular system. At the present time there are several methods used for this purpose and these involve the introduction into the vessels of chemical substances (gas bubbles, dyes, and labeled atoms) or special probes. Measurements made by these methods are very complex to carry out and do not guarantee sufficient accuracy.

The authors and their co-workers used the above method of measuring flow rate through the amplitude of the nuclear resonance signal to investigate blood flow. For this purpose they constructed an instrument for measuring the rate of blood flow by a fine method.

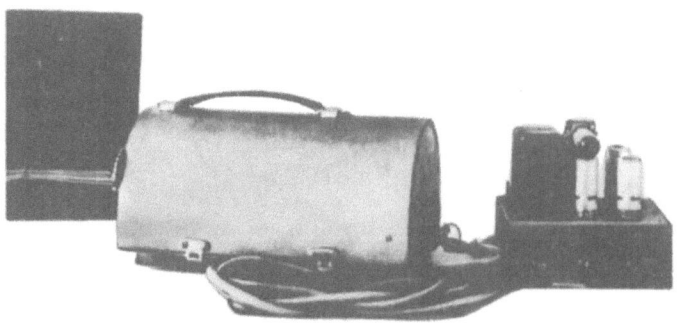

Fig. 2.7. Photograph of NMR amplitude flowmeter.

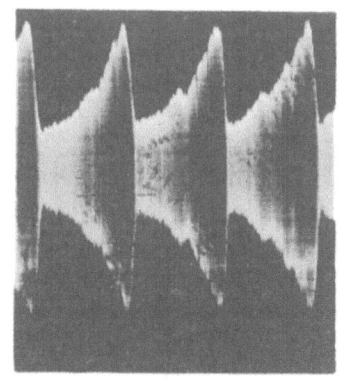

Fig. 3.7. Oscillogram of the signal from blood in the femoral artery of a dog.

A photograph of the apparatus is given in Fig. 2.7. To the left are glass inlet and outlet tubes, which are inserted into the lumens of the blood vessel. Under the cover of the instrument lie a system of magnets, radiofrequency coil of the detector, and the detection system tuned to a frequency of 12 MHz. The modulation frequency of the field is 1400 Hz. A nuclear resonance signal is observed with the same frequency. Its amplitude is proportional to the flow rate of the blood in the tube passing through the coil. The second unit lying to the right in the picture is designed for converting the signal from the first unit so as to record the amplitude on the recorder of an electrocardiograph.

An oscillogram of the amplified signal from the output of the first unit with the instrument connected to the femoral artery of a dog is given in Fig. 3.7. The signal amplitude changes at the pulse rate, which in the given case was one beat per second. The amplitude of the nuclear resonance signal from protons of blood is somewhat less than from protons of water, apparently because of the reduction in the Q-factor of the coil due to the considerable electrical conductivity of blood. This is the reason for the increased noise level.

The relaxation time T_1 was estimated from the rate of fall of the signal amplitude after a sudden stop in the blood flow (see Section 1.9). The fall in the signal amplitude in a time t is proportional to the factor exp $(-tA_\infty/T_1A_0)$, where A_∞ is the amplitude of the signal with a high blood flow rate equal to zero.

The measurements showed that the relaxation time of the protons of the blood of a dog T_1 equals 0.4 sec. The same value of T_1 was obtained for protons of mouse and human blood [40, 41].

CHAPTER 8

ABSOLUTE MEASUREMENT OF LIQUID FLOW RATE BY MAGNETIC LABELING OF NUCLEI

1.8. Principle of Measurement Method

The most direct method of determining absolutely the velocity of a liquid consists of measuring the time for the molecules of the liquid to cover a known distance. To use this method it is necessary to label the molecules or nuclei of the liquid in one section of the tube and record their arrival in another.

If the distance between these sections is l_0 and the time of travel of the molecules t_0, the mean velocity of the molecules of the liquid over the distance l_0, $W = l_0/t_0$. The nuclear magnetic resonance method makes it possible to label the nuclei of a liquid by polarizing them in a magnetic field. If there is a nuclear resonance detector at the end of the measuring section of the tube l_0, under certain definite conditions the signal amplitude will be proportional to the magnetization of the nuclei. A rapid change in the polarization of the liquid, produced at the beginning of the measuring section, produces a corresponding change in the amplitude of the nuclear resonance signal after a time t_0. By measuring t_0 it is possible to determine the liquid flow velocity.

The natural change in the polarization of nuclei occurs slowly with relaxation time $T = T_1$, which may be several seconds (T_1 is the natural spin-lattice relaxation time). For reducing T, the nuclei are labeled by artificial depolarization of the nuclei by means of a resonance oscillating field. For this purpose, before the liquid passes through the section l_0 it must be polarized with a strong field and a radiofrequency coil (nutation detector) placed at the beginning of this section. By exciting a resonance oscillating field in this coil it is possible to produce rapid depolarization or inversion of the magnetization of the liquid and by switching off this field, rapid polarization. The measurement may be made equally successfully with polarization and with inversion of the magnetization. A diagram of the measuring instrument is shown in Fig. 1.8 [38-42].

The liquid is polarized on passing through the strong field and when there is no potential on the coil of the nutation detector 3, it passes through the coil of the absorption detector 5 with a high magnetization, giving a nuclear resonance signal in the detector circuit 7. When a potential is applied to the coil 3, the magnetization of the nuclei in the liquid flowing through it is reduced to zero or reoriented relative to the direction of the external magnetic field (nutation occurs)and therefore the nuclear resonance signal disappears or has a phase opposite to the original signal. The measuring instrument may be operated in several ways. When polarization is used, at the beginning of the measurement the oscillating field is switched on and there is no absorption signal and at a moment of time t the oscillating field is switched off and simultaneously the time measuring system 8 is switched on automatically. At a moment of time $t + t_0$ there appears an absorption signal and the time measuring system is switched off so that its readings give the time t_0. When depolarization is used, at the beginning of the measurement the oscillating field is switched off and there is an absorption signal. At the moment of time t the oscillating field is switched on and at the same time the time measuring system. At a moment of time $t + t_0$ the absorption signal disappears and the time measuring system is switched off with a relay. When inversion of the magnetization is used, instead of the disappearance and appearance of the signal, the change in its phase is recorded.

2.8. Instrument Parameter Requirements

For the change in the nuclear resonance signal under the action of the resonance oscillating field H_{1n} in the coil of the nutation detector 3 (see Fig. 1.8) to be large enough, it is necessary to have the maximum possible signal amplitude in the absence of the field H_{1n} and the minimum possible amplitude in the presence of H_{1n}. The conditions for the maximum signal amplitude in the absorption detector were examined in Ch. 3 and when $T_{2n} \ll T_{1n}$, they have the form

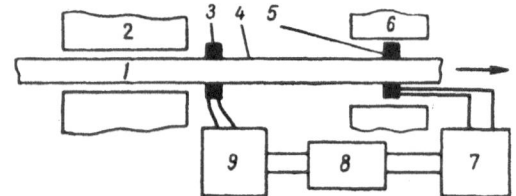

Fig. 1.8. Diagram of flowmeter using magnetic labeling of the nuclei: 1) space in polarizing field; 2) polarizing magnet; 3) coil of nutation detector; 4) measuring section of tube; 5) coil of absorption detector; 6) magnet of absorption detector; 7) NMR detector; 8) time measuring device; 9) radiofrequency oscillator.

$$Z_a \ll 1, \tag{1.8}$$

$$\frac{l_a}{2WT_{2n}} = \frac{l_a \gamma \Delta H_{\perp a}}{4W} \ll 1, \tag{2.8}$$

$$\frac{l_a}{W} \gamma H_{1a} \approx \frac{3\pi}{4}. \tag{3.8}$$

Here $Z_a = 1/(1 + \gamma^2 H_{1a}^2 T_{1n}^2 T_{2n}^2)$ is the saturation factor in the absorption detector, H_{1a} is half the strength of the resonance oscillating field in it, T_{1n} and T_{2n} are the effective transverse averaged across the flow and longitudinal relaxation times of the nuclear magnetization, l_a is the length of the absorption detector, and $\Delta H_{\perp a}$ is the nonuniformity of the external magnetic field directed at right angles to the stream in the absorption detector.

When these conditions hold and also $Z_a \ll l_a/WT_1$, from expression (5.3) it is found that the amplitude of the signal at the output of the NMR detector 7 is given by

$$A_a = 2.8\pi\eta_a Q_a N_a S_a \gamma H_a \left[M_{pa} + X_0 H_a \frac{0.4 l_a}{WT_1} \right] + A_{no}, \tag{4.8}$$

where M_{pa} is the magnetization of the nuclei entering the detector 5, H_a is the strength of the field in the volume of the detector 5, and A_{no} is the amplitude of the noise of the electronic circuit switching on the pickup.

If l_0 is the length of the measuring section of the tube 4 (see Fig. 1.8.) between the coils 3 and 5, lying in the field H_0, then the expression for M_{pa}, using expression (4.1) will be

$$M_{pa} = X_0 H_0 + (M_{z\,ex} - X_0 H_0) e^{-\frac{l_0}{WT_1}}, \tag{5.8}$$

where $M_{z\,ex}$ is the projection on the direction of the external field of the magnetization of the nuclei emerging from the nutation detector 3.

Natural relaxation in the volume of the nutation detector may be neglected and therefore, in the absence of the oscillating field H_{1n} in the coil 3

$$M_{z\,ex} = M_{en} \approx X_0 H_p (1 - e^{-\frac{l_p}{WT_1}}), \tag{6.8}$$

where M_{en} is the magnetization of the nuclei entering the nutation detector 3 and l_p is the length of the tube in the polarizer field H_p.

In this case the amplitude of the absorption signal

$$A_a = K \left[X_0 H_p (1 - e^{-\frac{l_p}{WT_1}}) e^{-\frac{l_0}{WT_1}} + X_0 H_0 \frac{0.4 l_a}{WT_1} + X_0 H_0 (1 - e^{-\frac{l_0}{WT_1}}) \right] + A_{no}, \tag{7.8}$$

$$K = 2.8\pi\eta_a Q_a N_a S_a \gamma H_a. \tag{8.8}$$

When there is a resonance oscillating field H_{1n} in the coil 3, the value of $M_{z\,ex} \ll M_{en}$ and the signal amplitude is reduced. As will be shown below, by selecting appropriate conditions it is possible to make $M_{z\,ex}$ sufficiently small and it may be neglected in expression (4.8). The corresponding minimum oscillation amplitude is the background:

$$A_b = K \left[X_0 H_0 \frac{0.4 l_a}{W T_1} + X_0 H_0 \left(1 - e^{-\frac{l_0}{W T_1}}\right) \right] + A_{no}. \tag{9.8}$$

The ratio of signal to background

$$\frac{A_a}{A_b} = \frac{X_0 H_p \left(1 - e^{-\frac{l_p}{W T_1}}\right) e^{-\frac{l_0}{W T_1}} + X_0 H_a \frac{0.4 l_0}{W T_1} + X_0 H_0 \left(1 - e^{-\frac{l_0}{W T_1}}\right) + \frac{A_{no}}{K}}{X_0 H_a \frac{0.4 l_a}{W T_1} + X_0 H_0 \left(1 - e^{-\frac{l_0}{W T_1}}\right) + \frac{A_{no}}{K}}. \tag{10.8}$$

For the instrument to operate reliably it is essential that $A_a/A_b > 10$. In this case from expression (10.8) we obtain the following condition:

$$X_0 H_0 \left(1 - e^{-\frac{l_0}{W T_1}}\right) + X_0 H_a \frac{0.4 l_a}{W T_1} + \frac{A_{no}}{K} < \frac{X_0 H_p e^{-\frac{l_0}{W T_1}} \left(1 - e^{\frac{l_p}{W T_1}}\right)}{9}. \tag{11.8}$$

The first and second terms in the left-hand part increase with a decrease in W. The third term is independent of W. In the right-hand part the first exponential factor predominates at small values of W, while the second predominates at large values of W. Thus, condition (11.8) is divided into two.

At the maximum velocity W_{max} (at the upper limit of the range of measurements) it has the form

$$\frac{A_{no}}{K} \leqslant \frac{X_0 H_p \left(1 - e^{\frac{l_p}{T_1 W_{max}}}\right)}{9}. \tag{12.8}$$

At the minimal velocity W_{min} (at the lower limit of the range of measurements) it has the form

$$X_0 H_0 \left(1 - e^{-\frac{l_0}{T_1 W_{min}}}\right) + X_0 H_a \frac{0.4 l_a}{T_1 W_{min}} + \frac{A_{no}}{K} \leqslant \frac{X_0 H_p e^{-\frac{l_0}{T_1 W_{min}}}}{9}. \tag{13.8}$$

From expression (12.8) we obtain the relation

$$1 - e^{-\frac{l_p}{T_1 W_{min}}} \geqslant \frac{9 A_{no}}{K X_0 H_p}, \tag{14.8}$$

where $K X_0 H_p$ is the amplitude of the signal at maximal polarization of the nuclei, i.e., when the conditions $l_p \gg T_1 W_{max}$ and $l_0 \ll T_1 W_{min}$ are fulfilled. Let us denote the signal-to-noise ratio at maximum polarization $K X_0 H_p/A_{no}$ by a_{max}. By rearranging relation (14.8) we obtain the condition

$$\frac{l_p}{T_1 W_{max}} \geqslant \ln \frac{a_{max}}{a_{max} - 9}. \tag{15.8}$$

In expression (13.8) the left-hand part represents the amplitude of the background, the three terms corresponding to the three different factors involved in the background. For an estimate it may be assumed that the contribution of each of the factors is the same and equal to 1/3 of the total background. On examining the contribution of the second term we obtain the condition

$$\frac{H_p}{H_a} \geqslant \frac{11 l_a e^{\frac{l_0}{T_1 W \text{min}}}}{T_1 W \text{min}} ; \tag{16.8}$$

on examining the contribution of the third term,

$$\frac{l_0}{T_1 W \text{min}} \leqslant \ln \frac{a_{\text{max}}}{27} ; \tag{17.8}$$

and that of the first term,

$$\frac{H_p}{H_0} \geqslant 27 \left(e^{\frac{l_0}{T_1 W \text{min}}} - 1 \right). \tag{18.8}$$

As will be shown subsequently, condition (18.8) is more than satisfied in practice and therefore the amplitude of the background is determined mainly by the second and third terms in the left-hand part of the inequality (13.8). This makes the conditions (16.8) and (17.8) less strict:

$$\frac{l_0}{T_1 W \text{min}} \leqslant \ln \frac{a_{\text{max}}}{18} , \tag{19.8}$$

$$\frac{H_p}{H_a} \geqslant \frac{7 l_a e^{\frac{l_0}{T_1 W \text{min}}}}{T_1 W \text{min}} . \tag{20.8}$$

Let us find the conditions with which the oscillating field H_{1n} in the coil 3 (see Fig. 1.8) ensures that $M_{z\,ex} \ll M_{en}$.

The change in M_z under the action of the resonance oscillating field is described by the expression obtained in appendix 1, which has the following form when $T_{2n} \ll T_{1n}$:

$$M_{z\,ex} = \left[\left(M_{en} - X_0 H_n Z_n \frac{T_{1n}}{T_1} \right) \left(\frac{e^{bt} + e^{-bt}}{2} + \frac{e^{bt} - e^{-bt}}{4bT_{2n}} \right) + \right.$$
$$\left. + \frac{X_0 H_n (1 - Z_n)}{bT_1} \cdot \frac{e^{bt} - e^{-bt}}{2} \right] e^{-\frac{t}{2T_{2n}}} + X_0 H_n Z_n \frac{T_{1n}}{T_1} , \tag{21.8}$$

where

$$b = \sqrt{\frac{1}{4T_{2n}^2} - \gamma^2 H_{1n}^2} ,$$

$Z_n \sim 1/(1 + \gamma^2 H_{1n}^2 T_{1n} T_{2n})$ is the saturation factor in the nutation detector, t is the time measured from the moment of appearance of H_{1n}, and H_n is the strength of the external field in the nutation detector.

This expression shows that with any values of b and t, $M_{z\,ex} > X_0 H_n Z_n T_{1n}/T_1$ and therefore, for $M_{z\,ex} \leq M_{en}$ the following condition must be fulfilled:

$$X_0 H_n Z_n \frac{T_{1n}}{T_1} \ll M_{en}. \qquad (22.8)$$

As the amplitude of the absorption signal is proportional to $M_{z\,ex}$ with an error of 10% when conditions (15.8) and (18.8) hold, the ratio of the signal to the background A_a/A_b equals the ratio $M_{en}/M_{z\,ex}$ with the same error and, consequently, $A_a/A_b < M_{en}T_1/X_0 H_n Z_n T_{1n}$. Therefore, for $A_a/A_b > 10$ in the nutation detector the following condition must be fulfilled:

$$Z_n < \frac{M_{en}T_1}{10 X_0 H_n T_{1\,n}}. \qquad (23.8)$$

In expression (21.8) the terms containing $X_0 H_n Z_n$ may be neglected:

$$M_{zex} = M_{en}\left(\frac{e^{bt}+e^{-bt}}{2}+\frac{e^{bt}-e^{-bt}}{4bT_{2n}}\right)e^{-\frac{t}{2T_{2n}}}. \qquad (24.8)$$

This relation is shown graphically in Fig. 2.8. The change in $M_{z\,ex}$ when $\gamma H_{1n}T_{2n} < 1.4$ has an oscillatory character and this is inconvenient as it results in the periodic appearance of a signal in the absorption detector. When $\gamma H_{1n}T_{2n} < 1.4$ $M_{z\,ex}$ changes aperiodically and with a decrease in $\gamma H_{1n}T_{2n}$, the time for $M_{z\,ex}$ to fall increases. As this time determines the speed with which the time measuring system is switched off and therefore is included in the error in the measurement of the time interval t_0, it is desirable to have the lowest possible value of T, and therefore the most favorable condition is

$$\gamma H_{1n}T_{2n} \approx 1.4, \qquad (25.8)$$

when from the graph in Fig. 2.8 we have

$$T = 3T_{2n} \approx \frac{6}{\gamma \Delta H_{\perp n}}, \qquad (26.8)$$

i.e., the rate of labeling of the liquid is determined by the transverse nonuniformity of the field in the nutation detector 3 (see Fig. 1.8).

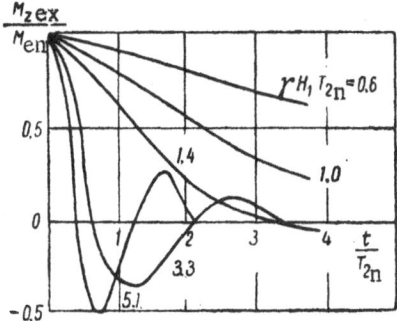

Fig. 2.8. Change in magnetization of nuclei under the action of the resonance oscillating field in the nutation detector.

3.8. Working Formulas for Measuring Liquid Velocity

Measurement of Maximum Average Velocity of Molecules. The time measuring system is switched on simultaneously with the input of high-frequency voltage to the coil 3 (see Fig. 1.8), when there is some delay τ_{on}, determined by the inertia of the relay of the switching system. In its turn, the appearance of the oscillating field in the coil 3 occurs with a time constant τ_{n1}, which is determined by its inductance, i.e., it also lags slightly behind the input of the voltage. From the moment that the oscillating field arises, the polarization of the liquid in the coil 3 begins to fall and disappears practically completely after a time $3T_{2n}$. By increasing the transverse nonuniformity of the external field in the coil 3 it is possible to make T_{2n} negligibly small so that the depolarization front arises practically instantaneously. At the initial moment it lies in the end section of the coil 3 closer to the coil 5. The switching on of the time measuring system lags behind the appearance of the depolarization front by $\tau_{on} - \tau_{n1}$.

The time measuring system is switched off as a result of a fall in the amplitude of the nuclear resonance signal produced by the arrival of the depolarized nuclei in the coil of the detector 5. This fall begins at the moment that the fastest molecules of the depolarization front enter the working volume of the detector. The distance l_0 between the nearest ends of the coils 3 and 5 is traversed by molecules with the maximum averaged velocity W_{max} in a time

$$t_{0min} = \frac{l_0}{W_{max}} . \tag{27.8}$$

If the instrument is constructed so that the relay switching off the time measuring device operates with the slightest decrease in the signal amplitude, the time measurement stops with the arrival of these fastest molecules of the depolarization front with a lag τ_{off}, caused by the inertia of the relay for switching off the time measuring system.

The time measured t_{min} differs from the true time of travel of the molceules t_{0min} by the value

$$\tau = \tau_{on} - \tau_{off} + \tau_{n1}. \tag{28.8}$$

and the maximum average velocity of the molecules

$$W_{max} = \frac{l_0}{t_{min} + \tau} . \tag{29.8}$$

Measurement of Minimum Average Velocity of Molecules. The molecules of the depolarization front with the minimum average velocity W_{min} are the last to arrive in the absorption detector 5. When they move out of the detector the change in the signal is complete. If a discriminator device provides for the operation of the relay for switching off the time measuring system when the amplitude of the nuclear resonance signal reaches its minimum value, the time measurement stops when the slowest molecules pass out of the detector. The time t_{max} measured in this way gives the minimum average velocity of the molecules

$$W_{min} = \frac{l_0 + l_a}{\sqrt{(t_{max} + \tau)^2 + \tau_d^2}} , \tag{30.8}$$

where l_a is the effective length of the coil of the absorption detector and τ_d is the time constant of the nuclear resonance detector.

Thus, by setting the discrimination level at the maximum or minimum value it is possible to measure the maximum or minimum average velocity of the molecules of the liquid in the flow.

Measurement of Mean Velocity of Liquid. At the moment that it arises, the depolarization front consists of a sharp boundary separating the liquid with the maximum M_{max} and minimum M_{min} magnetization of the nuclei and lies in the section of the tube at the end of the nutation coil 3 (see Fig. 1.8). This initial section moves with the mean velocity of the liquid W_{av} and during the process of moving the depolarization front spreads in both directions from the initial section so that during the time for traversing the measuring section its spread along the stream reaches the value l_d. To measure the mean velocity it is necessary to fix the moment that the initial section passes through the detector. For this it is necessary to know the relation of the amplitude of the nuclear resonance signal to the coordinate of the initial section inside the detector. As the signal amplitude depends on the magnetization of the nuclei, let us determine the mean magnetization of the nuclei through the volume of the detector at the moment that the initial section passes through the middle of the detector.

If all the molecules in the stream have a velocity close to W_{av}, then the depolarization front does not spread and at the moment that the initial section passes through the center of the absorption detector, the half of the volume of the detector in front of the initial section is occupied by liquid with a magnetization M_{max}, while the second half is occupied by liquid with magnetization M_{min}, i.e., the mean magnetization of the nuclei through the volume of the detector equals $(M_{max} + M_{min})/2$.

In actual fact, the molecules of the liquid have different velocities. Let us denote by N the total number of molecules in the volume of the detector and by dN(W) the number of molecules with an average velocity at a distance l_1 from the site of labeling of the liquid to the center of the detector in the range from W to W + dW.

Molecules with a velocity $W > W_{av}$ outstrip the initial section and at the moment of time considered, part of the volume of the detector in front of this section will be occupied by polarized liquid. Molecules with a velocity $W < W_{av}$ lag behind the initial section and part of the volume of the detector behind the section will be occupied by polarized liquid.

In this case the mean magnetization of the nuclei through the volume of the detector is determined by the following expression:

$$
\overline{M} = \frac{M_{0\,max}}{Nl_a}\left[\int_{W_2}^{\infty}\frac{dN}{dW}\int_{a}^{\beta}e^{-\frac{l_0}{WT_1}-\frac{x}{WT_1Z}}dW\cdot dx + \int_{W_1}^{W_2}\frac{dN}{dW}\int_{\gamma}^{\beta}e^{-\frac{l_0}{WT_1}-\frac{x}{WT_1Z}}dW\cdot dx\right]+
$$

$$
+\frac{M_{0\,min}}{Nl_a}\left[\int_{0}^{W_1}\frac{dN}{dW}\int_{a}^{\beta}e^{-\frac{l_0}{WT_1}-\frac{x}{WT_1Z}}dW\cdot dx + \int_{W_1}^{W_2}\frac{dN}{dW}\int_{a}^{\gamma}e^{-\frac{l_0}{WT_1}-\frac{x}{WT_1Z}}dW\cdot dx\right]
$$

$$
\alpha)\ x = 0;\ \ \beta)\ x = l_a;\ \ \gamma)\ x = \frac{W - W_{av}}{W_{av}}l_1 + \frac{l_a}{2}
$$

After integration with respect to x,

$$
\overline{M} = \frac{M_{max}}{N}\int_{W_2}^{\infty}\frac{dN}{dW}e^{a}\frac{\sin b}{b}dW + \frac{M_{min}}{N}\int_{0}^{W_1}\frac{dN}{dW}e^{a}\frac{\sin b}{b}dW +
$$

$$
+\frac{M_{max}}{N}\int_{W_1}^{W_2}\frac{dN}{dW}e^{a}\frac{e^{-c}+e^{-b}}{2b}dW + \frac{M_{min}}{N}\int_{W_1}^{W_2}\frac{dN}{dW}e^{a}\frac{e^{b}-e^{-c}}{2b}dW \tag{31.8}
$$

$$
W_1 = W_{av}\left(1 - \frac{l_a}{2l_1}\right),\ \ W_2 = W_{av}\left(1 + \frac{l_a}{2l_1}\right),\ \ a = \frac{W - W_{av}}{W\cdot W_{av}T_1Z}\left(l_0Z + \frac{l_a}{2}\right),\ \ b = \frac{l_a}{2WT_1Z},\ \ c = \frac{(W - W_{av})l_1}{W\cdot W_{av}T_1Z}.
$$

where Z is the saturation factor in the detector; l_0 is the distance from the site of labeling to the beginning of the detector: $l_0 = l_1 - l_a/2$; l_a is the length of the detector along the flow; M_{max} and M_{min} are the magnetizations at the center of the detector of nuclei of the unlabeled and labeled liquid, passing through the section l_1 with an average velocity W_{av}; and $M_{0\,max}$ and M_{0min} are the magnetizations of the liquid at the site of labeling. The cross section of the detector is assumed to equal that of the tube. In expression (31.8), dN/dW, which characterizes the distribution of the liquid with respect to average velocities in the stream, is unknown.

In the case of ideal laminar flow

$$\frac{dN}{dW} = \frac{dN}{dr} \cdot \frac{dr}{dW} ,$$ (32.8)

where r is the distance from the axis of the tube.

The value $(dN/dr) \, dr = 2\pi r n l_a dr$ equals the number of molecules in the absorption detector at a distance from the axis between r and $r + dr$ (n is the number of molecules in unit volume of the liquid). Hence

$$\frac{dN}{dr} = 2\pi r n l_a.$$ (33.8)

The distribution of the liquid velocity across a section of the tube with laminar flow has the form

$$W = 2W_{av} \left(1 - \frac{r^2}{R^2} \right) ,$$ (34.8)

where R is the radius of the tube.

From expression (34.8)

$$\frac{dr}{dW} = - \frac{R^2}{4r W_{av}} .$$ (35.8)

By substituting expressions (33.8) and (35.8) in formula (32.8) we obtain

$$\frac{dN}{dW} = - \frac{\pi l_a n R^2}{2W_{av}} = - \frac{N}{2W_{av}} .$$ (36.8)

By substituting equation (36.8) in formula (31.8), taking into account the fact that with laminar flow $W_{min} = 0$ and $W_{max} = 2W_{av}$ and assuming for simplicity that $a = b = c = 0$, we obtain

$$\overline{M} = \frac{M_{max} + M_{min}}{2} .$$ (37.8)

In actual fact, a, b, and c are not equal to zero. This introduces an error into the measurements which may be called the "relaxation error." In the case of laminar flow, this error may be quite considerable. It is determined by $l_0/W_{av}T_1$. If in the detector there are conditions with which the signal amplitude is proportional to the magnetization of the liquid flowing into the coil and is independent of its rate of entry, when the measuring system operates at the moment that the signal amplitude reaches half the maximum value, the relaxation error reaches 10% when $l_0/W_{av}T_1 = 0.1$, 18% when $l_0/W_{av}T_1 = 0.2$, 28% when $l_0/W_{av}T_1 = 0.4\%$, and 36% when $l_0/W_{av}T_1 = 0.8$. If in the detector there are conditions with which the signal amplitude is proportional to the magnetization of the nuclei and the rate of entry of the liquid bearing them, for measuring the average velocity when $l_0/W_{av}T_1 \ll 1$, the time measuring system must operate when the signal amplitude reaches 0.75 of the maximum value. In this case the relaxation error reaches 4% when $l_0/W_{av}T_1 = 0.1$, 12% when $l_0/W_{av}T_1 = 0.4$ and 20% when $l_0/W_{av}T_1 = 0.8$.

Because of the high value of the relaxation error in the measurement of the average velocity of laminar flow, in this case it is more advantageous to measure not the average velocity, but the maximum velocity, which equals twice the mean velocity, as the maximum velocity is measured without a relaxation error.

With turbulent flow of the liquid, if the size of the detector

$$l_a > (W_{max} - W_{min}) \frac{l_1}{W_{av}} , \quad a \ll 1, \quad b \ll 1, \quad c \ll 1$$

ιe expression for $\overline{M}$ has the form:

$$\overline{M} = \frac{M_{max} + M_{min}}{2} + \frac{M_{min} - M_{max}}{N W_{av}} \frac{l_1}{l_a} \int\limits_{W_{min}}^{W_{max}} (W - W_{av}) \frac{dN}{dW} dW. \qquad (38.8)$$

eliminating terms containing the factors a, b, and c).

The second term in this expression equals zero, i.e., in the first approximation

$$\overline{M} = \frac{M_{max} + M_{min}}{2}.$$

If the length of the detector is inadequate for it to accommodate the whole of the diffuse front of turbulent diffusion, to find the value of $\overline{M}$ it is necessary to know the law of the distribution of the molecules with respect to the velocities (dN/dW) in turbulent flow and also W_{max} and W_{min}.

As an investigation of turbulent diffusion showed, with clearly expressed turbulent flow of the liquid the maximum surge velocities, directed parallel and antiparallel to the movement of the stream, are approximately equal $|W_{max} - W_{av}| \approx |W_{av} - W_{min}|$. The value of dN/dW is independent of the sign of $W - W_{av}$, i.e., the distribution of the molecules with respect to surge velocities is independent of the direction in the first approximation.

Under these conditions and with a = b = c = 0, from expression (31.8) it follows that $\overline{M} = (M_{max} + M_{min})/2$. The fact that a, b, and c do not equal zero is the reason for the relaxation error $\Delta W_{av}/W$, which is much less with turbulent flow of the liquid than with laminar flow. A theoretical estimate shows that with $(\overline{\alpha}^2)^{1/2} \ll W_{av}$, it is determined by the expression

$$\frac{\Delta W_{av}}{W_{av}} = -\frac{l_0}{W_{av} T_1} \cdot \frac{d^2}{W_{av}^2}$$

where $(\overline{\alpha}^2)^{1/2}$ is the mean square deviation of the velocity of the liquid, averaged over the section l_0, from the average velocity

$$\overline{\alpha}^2 = \frac{1}{N} \int\limits_0^\infty (W - W_{av})^2 \frac{dN}{dW} dW.$$

The value of $\overline{\alpha}^2$ may be measured by the method described in Sections 7.9 and 1.10. In practice it falls with an increase in the length of the section l_0 and may be made sufficiently small. Let us estimate the effect on the amplitude of a change in the rate of exchange of energy of the nuclei with the coil over the length of the detector, assuming for simplicity that a = b = c = 0.

All the previous arguments hold if it is assumed that the distance l_a does not equal the length of the absorption detector and that this is the length of the actual initial section. Thus the magnetization of the nuclei averaged through the volume of the initial section $M_{av} = (M_{max} + M_{min})/2$, or if it is assumed that $M_{min} \ll M_{max}$ then $M_{av} = M_{max}/2$.

Let x be the coordinate measured from the beginning of the absorption detector along the stream, P(x) is the rate of exchange of energy of the nuclei with the coil, and x_0 is the coordinate of the position of the initial section in the detector. Let us examine the case when the length of the depolarization front $l_d \gg l_a$ and then the signal amplitude is related to the coordinate x_0 by the following expression:

$$\frac{A}{A_{max}} = \frac{\int\limits_0^{l_a} P(x) [M_{av} + \dot{M}(x - x_0)] dx}{\int\limits_0^{l_a} P(x) M_{max} dx} = \frac{M_{av}}{M_{max}} + \frac{\dot{M} \left[\int\limits_0^{l_a} (x - x_0) P(x) dx \right]}{\int\limits_0^{l_a} P(x) M_{max} dx}, \qquad (39.8)$$

where M is the gradient of the magnetization directed along the stream (when $l_a \ll l_d$, at a distance l_a it may be assumed to be independent of x). As $M_{av} = M_{max}/2$, then

$$\frac{A}{A_{max}} = \frac{1}{2} + \frac{\dot{M}}{M_{max}} \cdot \frac{\int_0^{l_a} (x - x_0) P(x)\, dx}{\int_0^{l_a} P(x)\, dx} . \tag{40.8}$$

From expression (40.8) it follows that $A = A_{max}/2$, when the initial section passes into the absorption detector to the coordinate $x_{0\frac{1}{2}}$ satisfying the condition

$$\int_0^{l_a} (x - x_{0\frac{1}{2}}) P(x)\, dx = 0. \tag{41.8}$$

If $P(x) = const$, expression (41.8) is simplified $l_a^2/2 - l_a x_{0\frac{1}{2}} = 0$, whence $x_{0\frac{1}{2}} = l_a/2$. This is also true if $P(x)$ has symmetry relative to the section with the coordinate $x = l_a/2$. Thus if $P(x) = const$ or has symmetry relative to the central section of the absorption detector, $A = A_{max}/2$ when the initial section has passed through to the section in the center of the absorption detector.

With low nonuniformity of the field in the detector

$$P(x) = \sin\left(\Theta \frac{x}{l_a}\right),$$

where $\Theta = \gamma H_{1a} l_a / W_{av}$ is the angle of nutation of the magnetization during the time for the nuclei to pass through the absorption detector. By substituting this value of $P(x)$ in expression (41.8), we obtain the condition determining $x_{0\frac{1}{2}}$

TABLE 1.8

Θ	0.4	1	$\frac{\pi}{2}$	$\frac{3\pi}{4}$	π	0.2	0
$\dfrac{x_{0\frac{1}{2}}}{l_a}$	0.67	0.65	0.64	0.62	0.5	0.66	0.67

TABLE 2.8

$\dfrac{l_a}{WT_1Z}$	0	0.2	0.3	0.5	1	∞
$\dfrac{x_{0\frac{1}{2}}}{l_a}$	0.5	0.48	0.46	0.45	0.42	0
$e^{-\frac{l_a}{WT_1Z}}$	1	0.82	0.74	0.61	0.37	0

$$\int\limits_0^{l_a} x \sin\left(\Theta\frac{x}{l_a}\right) dx = \int\limits_0^{l_a} x_{0\frac{1}{2}} \sin\left(\Theta\frac{x}{l_a}\right) dx,$$

whence

$$\frac{x_{0\frac{1}{2}}}{l_a} = \frac{(\sin\Theta)/\Theta - \cos\Theta}{1 - \cos\Theta}. \tag{42.8}$$

The relation of $x_{0\frac{1}{2}}/l_a$ to Θ is given in Table 1.8. When $\pi \geq \Theta \geq 0$, $x_{0\frac{1}{2}} = (0.5-0.7)\,l_a$, and when $\Theta = 3\pi/4$, $x_{0\frac{1}{2}} = 0.6 l_a$. With a high transverse nonuniformity of the external field in the detector, when the relaxation time $T_{2n} \ll l_a/W_{av}$ the rate of exchange of energy of the nuclei with the coil falls with the coordinate x according to the equation

$$P(x) = e^{-\frac{x}{WT_{1n}Z}},$$

where $Z = 1/(1 + \gamma^2 H_1^2 T_{1n} T_{2n})$ is the saturation factor. By substituting this value in expression (41.8), we obtain the condition determining $x_{0\frac{1}{2}}$,

$$x_{0\frac{1}{2}} \int\limits_0^{l_a} e^{-\frac{x}{WT_{1n}Z}} dx = \int\limits_0^{l_a} x e^{-\frac{x}{WT_{1n}Z}} dx.$$

After the integration it has the form

$$\frac{x_{0\frac{1}{2}}}{l_a} = \frac{WT_{1n}Z}{l_a}\left[1 + \frac{\frac{l_a}{WT_{1n}Z}e^{-\frac{l_a}{WT_{1n}Z}}}{e^{-\frac{l_a}{WT_{1n}Z}}-1}\right]. \tag{43.8}$$

This relation is given in Table 2.8.

It is possible to determine the value of $l_a/WT_{1n}Z$ in the nuclear resonance detector by measuring the ratio of the vectors of the magnetization of the nuclei in the stream of liquid at the exit and entrance to the detector. The ratio of these values equals $\exp(-l_a/WT_{1n}2)$.

In practice, when autodyne detectors are used $l_a/WT_{1n}Z < 0.5$, i.e., $0.5 \geq x_{0\frac{1}{2}}/l_a \geq 0.45$.

Let us examine the case when $l_d \ll l_a$. Between the polarized and unpolarized liquids there is a sharp boundary, lying in the initial section and moving together with it. When the initial section passes through the absorption detector, the relation of the signal amplitude to x_0 (the distance of the initial section from the front end of the coil in Fig. 1.8) is

$$\frac{A}{A_{max}} = \frac{\int\limits_{x_0}^{l_a} l(x)\,dx}{\int\limits_0^{l_a} P(x)\,dx}. \tag{44.8}$$

If $P(x) = const$, then from expression (44.8) we find that $A/A_{max} = (l_a - x_0)/l_a$; when $A/A_{max} = 1/2$ the coordinate of the initial section $x_{0\frac{1}{2}} = 0.5$. This is also true if $P(x) = P(l_a - x)$.

TABLE 3.8

Θ	0.4	1	$\dfrac{\pi}{2}$	$\dfrac{3\pi}{4}$	π	0.2	0.1	0
$\dfrac{x_{0\frac{1}{2}}}{l_a}$	0.70	0.69	0.66	0.6	0.5	0.7	0.71	0.71

TABLE 4.8

$\dfrac{l_a}{WT_{\ln}Z}$	0	0.2	0.3	0.5	1
$\dfrac{x_{0\frac{1}{2}}}{l_a}$	0.5	0.48	0.46	0.44	0.38
$e^{-\dfrac{l_a}{WT_{\ln}Z}}$	1	0.82	0.47	0.61	0.37

When $P(x) = \sin(\Theta x/l_a)$, from expression (44.8) we find that

$$\frac{A}{A_{\max}} = \frac{\cos\left(\Theta\,\dfrac{x}{l_a}\right)}{\cos\Theta - 1} \,, \tag{45.8}$$

whence when $A/A_{\max} = 1/2$.

$$\frac{\cos\Theta - \cos\left(\Theta\,\dfrac{x_{0\frac{1}{2}}}{l_a}\right)}{\cos\Theta - 1} = \frac{1}{2} \,. \tag{46.8}$$

This relation is given in Table 3.8. When $\pi \geq \Theta \geq 0$, $x_{0\frac{1}{2}} = (0.5 - 0.7)\,l_a$; and when $\Theta = 3\pi/4, x_{0\frac{1}{2}} = 0.6 l_a$. When $P(x) \approx \exp(-x/WT_{\ln}Z)$, from expression (43.8) we find that

$$\frac{e^{-\dfrac{x}{WT_{\ln}Z}} - e^{-\dfrac{x_{0\frac{1}{2}}}{WT_{\ln}Z}}}{e^{-\dfrac{l_a}{WT_{\ln}Z}} - 1} = \frac{1}{2}$$

or

$$\frac{x_{0\frac{1}{2}}}{l_a} = \frac{WT_{\ln}Z}{l_a}\ln\frac{2}{e^{-\dfrac{l_a}{WT_{\ln}Z}} + 1} \,. \tag{47.8}$$

This relation is given in Table 4.8, when $0 < l_a/WT_{\ln}Z \leq 0.5, 0.5 \geq x_{0\frac{1}{2}}/l_a \geq 0.44$.

To summarize it may be stated that for measuring the average velocity the discrimination level should be set at half the maximum amplitude as only in this case is the result of the measurements independent of the effect of turbulent diffusion. Then

$$W_{av} = \frac{l_0 + x_{0\frac{1}{2}}}{\sqrt{(t+\tau)^2 + \tau_d^2}}.$$ (48.8)

If $P(x) = const$ or $P(x) = P(l_a - x)$, the time measuring system will be switched off when the initial section of the depolarization front reaches the center of the detector, i.e., in this case $x_{0\frac{1}{2}} = 0.5l_a$ and the average velocity of the liquid flow may be determined from the formula

$$W_{av} = \frac{l_0 + 0.5l_a}{\sqrt{(t+\tau)^2 + \tau_d^2}}.$$ (49.8)

If in the detector there is a high transverse nonuniformity of the field, i.e., $T_{2n} \ll l_a$, then at low liquid velocities, when $l_a/WT_{1n}Z = 1$, $x_{0\frac{1}{2}}/l_a \approx 0.4$, and at high velocities, when $l_a/WT_{1n}Z \ll 1$, $x_{0\frac{1}{2}}/l_a = 0.5$.

If the transverse nonuniformity of the field in the detector is not too great, i.e., $T_{2n} > l_a/W_{max}$, then at high velocities, when $(l_a/W_{max})\gamma H_1 \ll 1$, $x_{0\frac{1}{2}}/l_a \approx 0.7$. When the velocity decreases this ratio falls. When $\gamma H_1 l_a/W \approx \pi$, $x_{0\frac{1}{2}}/l_a \approx 0.5$.

Thus, at high liquid velocities when $l_a/W < T_{2n}$ and $0 < \gamma H_1 l_a/W < 1$, the value of $x_{0\frac{1}{2}}/l_a \approx 0.7$, and at low velocities, when $l_a/W > T_{2n}$ and $l_a(1 + \gamma^2 H_1^2 T_{1n} T_{2n})/WT_{1n} \approx 1$, the value of $x_{0\frac{1}{2}}/l_a \approx 0.4$, if we are limited to velocities at which $0 < l_a/WT_{1n} < 0.1$.

The indeterminacy of the value of $x_{0\frac{1}{2}}$, which is approximately $0.1l_a$, produces a relative error in the absolute measurement of the flow rate

$$\frac{\Delta W}{W} \approx \frac{0.1l_a}{l_0 + 0.5l_a}.$$ (50.8)

The length of the detector approximately equals the diameter of the tube d and then

$$\frac{\Delta W}{W} = \frac{0.1d}{l_0 + 0.5d}.$$ (51.8)

Thus, for reducing the error it is necessary to increase the length of the measuring section.

All the discussions in the present section refer to work with depolarization of the liquid. At the initial moment of time after the oscillating field has been switched on, the tube 4 in Fig. 1.8 is filled with polarized liquid and the space inside the coil 3 is filled with depolarized liquid. Therefore, the depolarization front passes through the edge of the coil 3 nearest to the coil 5.

In the work with polarization, when in the initial moment the resonance oscillating field of the coil 3 is switched off, the polarization front lies close to the edge of the coil 3 furthest from the coil 5 as the tube 4 and the space inside the coil are filled with depolarized liquid. There are no further fundamental differences between the two methods and therefore the results obtained are valid for work with polarization when in all the expressions l_0 is replaced by $l_0 + l_n$, where l_n is the effective length of the coil 2 and with a corresponding change in the discrimination level.

4.8. Method of Determining Effective Length of Coil

In the measurement of the maximum velocity of molecules of a liquid with depolarization, from expression (29.8)

$$W_{max} = \frac{l_0}{t_{dep} + \tau} ,$$

where t_{dep} is the time determined by the time measuring system with the maximum discrimination level. In the measurement of the same velocity with polarization

$$W_{max} = \frac{l_0 + l_n}{t_{pol} + \tau} ,$$

where t_{pol} is the time determined by the measuring system with the minimum discrimination level. By equating these two expressions we obtain the value of l_n

$$l_n = \frac{l_0 (t_{pol} - t_{dep})}{t_{pol} + \tau} . \tag{52.8}$$

This effective length is not essentially equal to the geometric length of the coil along the stream. The value $l_n/2$ gives the distance from the center of the coil to the section where the depolarization or polarization front lies at the initial moment. The distance l_0 must be measured from precisely this section. If the coil 3 is geometrically similar to the coil 5 (see Fig. 1.8), their effective lengths are approximately the same ($l_n \approx l_a$) and then

$$l_0 = l_c - l_{n'} \tag{53.8}$$

where l_c is the distance between the centers of coils 3 and 5.

5.8. Method of Determining Corrections τ and τ_d

The corrections τ and τ_d may be found by summing their separate components, which may be calculated or measured. This method is not reliable as in actual fact the individual lags cannot be summed but are superposed on each other. The most reliable method of determining the corrections is direct experimental measurement, which may be done, for example, by measuring the same liquid velocity with two values of the distance l_0, namely, l_0' and l_0''. From expression (29.8) we obtain

$$W_{max} = \frac{l_0'}{t_{min}' + \tau} = \frac{l_0''}{t_{min}'' + \tau} ,$$

and from expression (30.8)

$$W_{min}^2 = \frac{(l_0' + l_a)^2}{(t_{max}' + \tau)^2 - \tau_d^2} = \frac{(l_0'' + l_a)^2}{(t_{max}'' + \tau)^2 - \tau_d^2} ,$$

whence we obtain expressions for τ and τ_d:

$$\tau = \frac{l_0' t_{min}'' - l_0'' t_{min}'}{l_0'' - l_0'} , \tag{54.8}$$

112

$$\tau_d^2 = \frac{(l_0'' + l_a)^2 (t'_{max} + \tau)^2 - (l_0' + l_a)^2 (t''_{max} + \tau)^2}{(l_0'' + l_a)^2 - (l_0' + l_a)^2} . \tag{55.8}$$

The correction τ may be made insignificant by using fast trigger devices in the measuring system. The correction τ_d is connected with the inertia of the nuclear resonance detector. In the observation of a signal in a weak magnetic field with a detector of the autodyne type it is determined mainly by the regenerated quality factor of the receiver coil. When other types of detector are used and when the signal is observed in a strong field the lag is determined by the band of the signal amplification channel.

6.8. Example of Practical Investigation of Flowmeter

The practical measurement of a liquid velocity was carried out with the experimental apparatus whose block diagram is shown in Fig. 1.8. An iron-clad magnet with an interpolar space of 400 cc was used for preliminary polarization of the liquid. The diameter of the tube (connecting tube) was 0.45 cm. The absorption detector was a hollow glass cylinder, 18 mm in diameter, the coil was found in two layers with PÉ -0.15 wire, the length of the coil was 10 mm, and the volume of the space inside the coil was 2.5 cc. This large working volume was used to increase the effect of x_0. The effective length of the working volume (assuming that its cross section equaled the cross section of the tube) l_a = 15 cm, x_0 = 0.5 l_a being comparable with l_0 and readily estimated from the experimental results. The construction of the nutation detector 3 was similar to that shown in Fig. 3a.3. The coil, 4 mm in length, was wound directly on the tube. The strength of the external field in it was about 1 oe. To the coil 3 was connected a ZG-10 audiofrequency oscillator, tuned to a frequency of 4550 Hz. The coil 5, which was connected into the circuit of an autodyne nuclear resonance detector, was placed in the stray field of the polarizing magnet with a strength of about 12 oe, modulated with a frequency of 50 Hz due to the pulsation of the voltage supply.

The signal was observed in the form of a sinusoidal potential with a frequency of 50 Hz. This potential was converted into positive pulses with a frequency of 50 Hz, whose amplitude was proportional to the amplitude of the nuclear resonance signal. These pulses were fed through a discriminator device to a PS-64 scaler. The scaler was switched on with a tumbler switch, which simultaneously connected the output of the ZG-10 to the coil 3. From the moment that the tumbler was switched on the nuclear resonance signal pulses were counted until there was a fall in the amplitude of the pulses when demagnetized liquid entered the detector. With the discrimination set at the maximum level, counting stopped with a very slight decrease in the signal amplitude, when it was set at half the pulse amplitude, counting stopped when the signal fell to a half, etc. The time measured by the system was determined through the number of pulses counted n:

$$t = \frac{n}{50} .$$

The experimental relation of the time determined by the measuring system to the discriminator bias set is shown in Fig. 3a.8. With the maximum bias U_{max} the pulse count stopped at a signal amplitude which differed from the maximum by very little. The points of group I correspond to $l_0 = l_0'$ = 10 cm and those of group II to $l_0 = l_0''$ = 114.5 cm.

The following values were determined directly from the graph:

$$t'_{min} = 0.16_{sec}$$

$$t'_{max} = 0.24_{sec}$$

$$t'_{min} = 0.62_{sec}$$

$$t'_{max} = 0.74_{sec}$$

They make it possible to determine the corrections τ and τ_d. From formula (54.8), τ = −0.115 sec and from formula (55.8), τ_d = 0.04 sec.

As an example, Fig. 3b.8 gives the results of an analogous experiment with the time constant of the detector circuit increased artificially by reduction of the transmission band of the nuclear resonance signal amplifier. In this case t'_{min} = 0.16 sec, t'_{max} = 0.3 sec, t''_{min} = 0.62 sec, and t''_{max} = 0.76 sec.

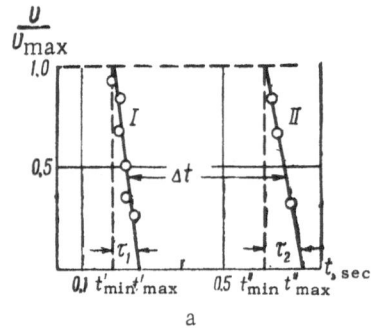

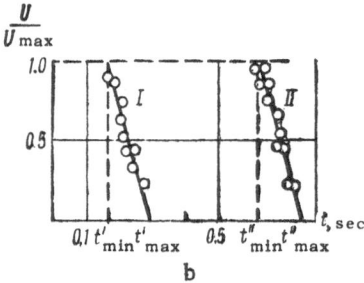

Fig. 3.8. Relation of the time determined by the time measuring system in the flowmeter to the discrimination level set: a) with a short time constant of the NMR detector; b) with an increased time constant of the NMR detector.

In this case, from formula (54.8) we find that $\tau = -0.115$ sec and from formula (55.8), $\tau_d = 0.17$ sec.

The experimental results given in Fig. 3.8 make it possible to determine the discrimination level at which the average velocity of the molecules is measured and x_0 in the expression for the mean velocity (48.8): for case I, $W_{av} = (l_0' + x_0)/\sqrt{(t' + \tau)^2 - \tau_d^2}$, and for case II, $W_{av} = (l_0'' + x_0)/\sqrt{(t'' + \tau)^2 - \tau_d^2}$. As x_0, τ, and W_{av} are the same in both cases with the same discrimination level set, and $\tau_d^2 \ll (t + \tau)^2$, then $W_{av} = (l_0'' - l_0')/(t'' - t')$.

The value of W_{av} measured independently by a volumetric method was 219 cm/sec, $l_0'' - l_0' = 104.5$ cm, and then $t'' - t' = 0.478$ sec. From Fig. 3a.8 we find that this value corresponds to a discrimination level $U = U_{max}/2$ and knowing the discrimination level it is possible to find the value of t from the data in Fig. 3a.8. From this value it is possible to determine x_0 from the formula $x_0 = W_{av}(t + \tau) - l_0$, which follows from the expression for the mean velocity (48.8). In case I ($t' = 0.2$ sec) $x_0 = 7.95$ cm; and in case II ($t'' = 0.78$ sec) $x_0 = 8.5$ cm.

The effective length of the detector $l_a = 15$ cm and therefore in case I, $x = 0.53\, l_a$ and in case II, $x_0 = 0.56\, l_a$, which correspond to the theoretical limits of the change in x_0. The experimental apparatus described made it possible to check the working formulas, but it did not give a high accuracy for the measurements as the time measurement error was 0.02 sec because of the low frequency of the pulses. The error of the method was estimated on another apparatus. In this apparatus we used a polarizing magnet with an interpolar space of 300 cc and a strength of about 5000 oe and a secondary magnet with a volume of 20 cc and a strength of 2800 oe. The diameter of the tube was 4.5 mm and the length of the measuring section 146 mm. The first coil was similar to that used in the previous experiment, while the second coil contained eight turns of wire wound in one layer. The volume of the absorption detector was about 0.2 cc. The signal was recorded with an autodyne detector and the maximum modulation frequency of the field was 2000 Hz, which made it possible in principle for the error in the time measurement to equal 0.0005 sec.

The results of the measurements at several liquid velocities W are given in Table 5.8, where f is the repetition frequency of the signal, and n is the number of pulses counted by the scaler. For each velocity we give three series of values of n, each of which consisted of ten separate measurements and the result of the series was determined as the arithmetic mean (in further refinement of the instrument it is proposed to do this automatically).

The relative error, determined from the reproducibility of the number of pulses counted, did not exceed 0.25% and at a low flow rate it fell to 0.1%. In addition to this value, the error in the velocity measurement also includes the error in the determination of the length of the section l_0, which is $0.1\,(l_d + l_a)/l_a$, where d is the diameter of the detector coil and l_a is the length of the absorption detector, and the error in the frequency measurement. Both of these errors are less than 0.1%, i.e., the error in the velocity measurement is determined by the scatter in the values of n and in the apparatus constructed this did not exceed 0.25%.

This statistical scatter arises for three reasons: a) electronic interference, 2) pulsations in the flow rate given by the pump, and 3) the instability of the discrimination level of the scaler. The first two reasons cannot be estimated. The effect of the third factor was estimated experimentally.

As was established, the average velocity is measured if the pulse count stops at the moment when the signal amplitude corresponds to half the maximum, i.e., if the discrimination level $U = A_{max}/2$. At the moment that the pulse count stops there passes through the center of the detector the section of the depolarization front, which has passed through the length l_0 with an average velocity of W_{av}. If the relative shift of the discrimination level $\Delta U/U = \pm 1$, i.e., if the maximum or minimum discrimination level is set, the pulse count stops when this section is at a distance $x = \pm 1/2(l_d + l_a)$, from the center of the detector, where l_d is the length of the polarization front resulting from turbulent diffusion. According to experimental data, x depends linearly on front i.e., if the discrimination level is displaced from the value of $U = A_{max}/2$ by ΔU, at the moment that counting stops this section is at a distance $x = 0.5(\Delta U/U)(l_d + l_a)$ from the center of the detector. This gives an error in the measurement of the average velocity $\Delta W_{av} = -0.5(\Delta U/U) \cdot (l_d + l_a)/(t + \tau)$, where t is the time determined by the system. As the length of the detector is much less than the length of the measuring section, then $W_{av} = l_0/(t + \tau)$ and the relative error in the velocity measurement

$$\frac{\Delta W_{av}}{W_{av}} = -\frac{\Delta U}{U} \cdot \frac{l_d + l_a}{2l_0}.$$

In the experimental apparatus $l_0 = 146$ cm, $l_a = 15$ cm; and l_d was determined with $W_{av} = 427$ cm/sec and equaled 12 cm, i.e., $\Delta W_{av}/W_{av} < 0.1 \Delta U/U$. From this it follows that with a relative shift in the discrimination level of 1%, there appears an error in the velocity measurement of 0.1%, i.e., the discrimination level must be kept equal to half the signal amplitude with an accuracy less by a factor of 10 than the specific accuracy of the liquid flow rate measurement. To estimate the accuracy of absolute measurements, the flow rate was determined simultaneously with an error of 1% from the time to fill a calibrated volume. No deviations were observed within the limits of this error.

Flow rate may be measured automatically be several methods. In the pulse-frequency method a potential of resonance frequency is fed to the demagnetizating coil through an electronic switch, which is open when there is a signal at the output of the nuclear resonance detector circuit and closed when there is no signal.

In this system the nuclear resonance signal appears periodically with a frequency proportional to the liquid flow rate. An oscillogram of such a nuclear resonance signal with a constant flow rate is shown in Fig. 4a.8 and with a varying flow rate, in Fig. 4b.8. The signal consists of a sinusoidal oscillation with a frequency of 1500 Hz, whose individual periods are indistinguishable on the photograph. The frequency of the modulating oscillations equals W/l Hz, where W is the liquid velocity and l the length of the measuring section.

TABLE 5.8.

W, cm/sec	427	172	110	72
f, Hz	2000	500	500	500
n_1	686.6	416.7	666.1	1020
n_2	678.6	416.3	663.6	1017.6
n_3	683.8	418.7	661.4	1017.8
$\bar{n}$	683.0	416.9	663.7	1018.5
$\dfrac{\Delta n}{\bar{n}}$	0.21	0.24	0.24	0.1

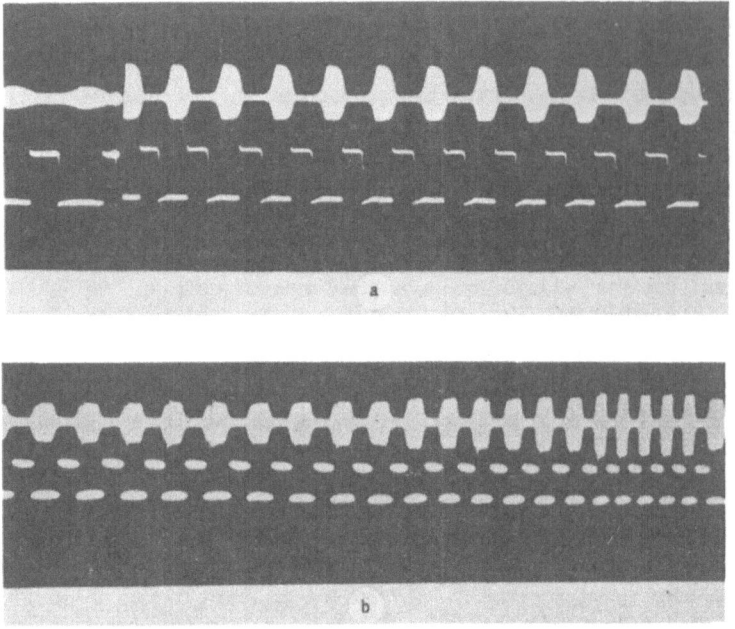

Fig. 4.8. Oscillograms of the nuclear resonance signal in a flow detector during the measurement of a liquid velocity of the pulse-frequency method: a) constant flow rate; b) varying flow rate.

Fig. 5.8. Frequency-pulse flowmeter.

In the time-pulse method the resonance potential is fed to the demagnetizing coil in short pulses, repeating at some constant frequency. Brief falls in the nuclear resonance signal amplitude occur with the same frequency and these are readily converted into recordable pulses of any polarity. The liquid velocity may be determined from the phase shift between the pulses of demagnetizing potential and the pulses recorded.

Both methods require automatic fixation of the moment when the signal amplitude in the absorption detector reaches half the maximum value, which changes with a change in the liquid flow rate, its chemical composition, and the sensitivity of the detector circuit. For this purpose it is possible to use a follow-up discriminator, whose level is automatically set at half the maximum amplitude of the signal in the detector. The discrimination level should not respond to the periodic disappearance of the signal, i.e., the discriminator should have quite a high inertia, which increases the inertia of the flowmeter.

Another possibility is the use of a discriminator with a constant level and automatic regulation of the signal amplitude amplification before the discriminator.

The simplest automatic regulation of the discrimination level may be achieved by using inversion of the magnetization for labeling the liquid. In this case, on one side of the depolarization front will be liquid with a nuclear magnetic moment M_{max} and on the other side, liquid with a magnetic moment $M_{min} = -M_{max}$. When the initial section passes through the center of the absorption detector the latter will contain liquid with an average magnetization through the volume $M = M_{max}/2 + M_{min}/2 = 0$.

Thus, in this case, regardless of the magnitude of the magnetization of the nuclei in the liquid or the sensitivity of the detector circuit, for measuring the velocity it is necessary to fix the moment at which the amplitude of the signal in the detector passes through zero. Without increasing the lag of the measurements, this method doubles the signal-to-noise ratio.

As was shown in Ch. 2, for inversion of the magnetization of the nuclei in a flowing liquid, the gradient of the external magnetic field grad H in the nutation coil should be along the liquid flow and the following condition should be fulfilled:

$$\gamma H_1^2 \geqslant 3W \operatorname{grad} H, \qquad (45.2)$$

where H_1 is half the strength of the resonance oscillating field in the coil. For demagnetization the gradient of the external field should be across the liquid flow.

The initial extent of the depolarization front $\Delta l = 2H_1/(\operatorname{grad} H)$, taking into account condition (56.8) will be $\Delta l = 6W/\gamma H_1$, i.e., with a sufficiently high strength of the oscillating field, Δl, as when demagnetization of the liquid is used, will be negligibly small.

The length of the measuring section is determined by the distance between the initial position of the depolarization front in the nutation coil and the center of the absorption detector. When the liquid is labeled by inversion of the magnetization of the nuclei the initial position of the depolarization front may be set at any section of the stream in the nutation coil by selecting the frequency of the oscillating field corresponding to the precession frequency of the nuclei in this section. With random changes in the strength of the external magnetic field or in the frequency of the oscillating field, the position of the section in the stream will move slightly along it, but if the gradient of the external field is high enough, this movement will be negligibly small.

Figure 5.8 gives a photograph of the first NMR frequency-pulse flowmeter with a range of 3-50 liters/hr, built under the direction of the authors.

The external unit in the cylindrical housing contains the magnet system and two radiofrequency coils, connected by a six-meter cable to the input plug of the electronic unit. In another form of the instrument, the cable has a length of 120 m.

The tube in which the liquid flow rate is measured passes through the external unit along its axis. Information on the flow rate is converted into a direct-current voltage and measured with a needle instrument on the front panel of the electronic unit or with a recording voltmeter connected to the terminals. The first pulse-phase NMR flowmeter, which was built by A. I. Zhernovoi and V. M Stasevich, had a range of 0.5-4 m^3/hr. The error of both instruments was about 1%.

CHAPTER 9

MEASUREMENT OF LONG RELAXATION TIMES T_1
IN A CONTINUOUS STREAM OF LIQUID

1.9 Review of Methods of Measuring T_1

Direct Method [169, 170]. This is normally used if $T_1 > 0.5$ sec. For measurements by this method a nuclear resonance signal is obtained at low saturation and then the signal is saturated until it practically disappears by increasing the amplitude of the resonance oscillating field. At the moment t_0 the oscillating field is reduced to its previous value and then the amplitude of the signal A begins to increase with a relaxation time T_1 in accordance with the equation

$$A = A_\infty (1 - e^{\frac{t-t_0}{T_1 Z}}),$$

where A_∞ is the signal amplitude when $t - t_0 \gg T_1 Z$; and Z is the saturation factor. The relation of $\ln[A_\infty/(A_\infty - A)]$ to $t - t_0$ is plotted and $T_1 Z$ found from the slope of the line obtained. As the saturation is low, $Z \approx 1$, i.e., $T_1 Z \approx T_1$. Motion picture filming is often used to record the change in the signal amplitude with time.

A similar method was used in the apparatus designed by Polish scientists [171, 172]. The signal was observed with modulation of the field with a frequency of 50 Hz. Total saturation of the signal is produced by increasing the amplitude of the resonance oscillating field. At some moment of time the external field changes sharply and the conditions for resonance are disturbed and the magnetization of the nuclei begins to increase with a relaxation time T_1. At the same moment, an electrolytic potentiometer is switched on and this smoothly adjusts the external field to the resonance value in a given time Δt. At the moment of resonance, a signal is observed with an amplitude A, proportional to the magnetization of the nuclei, and this is photographed. By constructing the relation $\ln[A_\infty/(A_\infty - A)]$ to Δt it is possible to determine T_1 from the slope of the line. In this apparatus the signal was observed with a large amplitude of the oscillating field, but the saturation did not affect the relaxation process as the latter did not proceed in the resonance field. The accuracy of the measurements was 3-5%.

Pulsed Methods. In Torrey's method [72, 73, 79, 80, 173] the magnetization of the nuclei in the sample is saturated with a short powerful pulse of the resonance oscillating field. A time Δt after the removal of the saturating pulse a second pulse is applied and a nuclear resonance signal A is observed. T_1 is found by plotting $\ln[A_\infty/(A_\infty - A)]$ against Δt (A_∞ is the signal amplitude when $\Delta t \gg T_1$).

In Hahn's spin echo method [68-71, 174-179] the sample is subjected to a series of pulses of the resonance oscillating field. Measurement of the relation of the amplitude of the secondary signals appearing to the time interval between the primary pulses makes it possible to determine T_1.

Low-Frequency Modulation Method. This method consists of measuring the signal amplitude at various modulation frequencies of the external field, comparable with $1/T_1$ [130, 180-182] or of measuring two alternating signal amplitudes with unsymmetrical modulation of the field. The error may be estimated from the results of Chiarotti and Guilotto's work, in which they measured T_1 of water saturated with oxygen at a pressure of 1 atm (1.4 ± 0.2 sec) and T_1 of deoxygenated water (3.6 ± 0.2 sec), and it is 6-14%.

None of the methods examined is suitable for measuring T_1 of a moving liquid if l/W (l is the length of the detector and W is the velocity of the liquid) is comparable with T_1 as the inflow of polarized liquid into the detector reduces the effective measured relaxation time, while the inflow of unpolarized liquid increases it. This produces a relative error in the measurements of WT_1/l.

Suryan's Method [1]. This method is suitable for measuring T_1 in the range of 0.01-0.1 sec. It uses the relation of the signal amplitude A to the velocity of the liquid W through the detector. By constructing a graph of the relation of $(A-A_0)/A_0$ (A_0 is the amplitude of the signal in the stationary liquid) to W/l (l is the length of the detector coil), it is possible to determine the relaxation time.

Hrynkiewicz and Waluga's Method [8]. In their apparatus the measurement process consists of plotting the relation of the signal amplitude in a flow detector to the liquid flow velocity. For determining T_1 it is necessary to construct a graph of this relation and extrapolate it to zero and infinite flow rates. Suryan proposed that the measurements should be carried out with deliberate high saturation, while Hrynkiewicz and Waluga proposed that the measurements should be carried out with low saturation and the saturation factor determined from the graph. This made it possible to raise the upper limit of the measurements from 0.1 sec to several seconds. The examples presented by the authors show that the measurement error for T_1 = 2.7 sec is 10% and for T_1 = 0.15 sec, 80%.

Antonowicz and Gaussen's Method. These authors proposed a method of measuring the relaxation time T_1 by periodic stopping of the liquid [6, 9, 10]. As was mentioned in the introduction, Antonowicz's arguments are valid on condition that there is complete mixing of the liquid in the detector in a time which is much less than the time for the nuclei to pass through the detector, which is impossible to achieve and therefore his method is only suitable for an approximate estimate of T_1. Gaussen's method has no particular advantages over the normal direct method of measuring T_1 when the increase in the signal is recorded on a motion picture film after removal of the nuclear resonance saturation with a strong oscillating field.

Zhernovoi and Pivovarov's Method. Superficially this is similar to Gaussen's method. The liquid passes through the detector at a high speed and then it is stopped and the change in the signal amplitude with time recorded. In contrast to Gaussen's method, in this method the liquid, before passing through the detector, passes through a polarization vessel lying in the same magnetic field H, i.e., polarized liquid enters the detector and the amplitude of the signal A in the flowing liquid is not equal to zero, but is proportional to the vector of the magnetization of the nuclei carried into the detector by the liquid. When the liquid is stopped, the signal amplitude falls with a relaxation time T_1 to a value $A_1 = AZ$. Thus, in this method it is possible to determine Z from the ratio of the amplitudes of the signals in the stationary and moving liquids in addition to T_1Z. This makes it possible to make the measurements with the optimal value Z = 0.5, while in Gaussen's method and in the direct method, where Z is unknown, it is necessary to work with $Z \approx 1$ and this reduces the signal-to-noise ratio. This method was used in practice to measure the relaxation time T_1 of blood in a living dog, which was found to equal 0.4 sec.

There is yet another method of measuring the relaxation time by determining the rate of change of the magnetization of the nuclei during the flow of the liquid from the polarizer into the nuclear resonance detector [46-48]. This method of measuring T_1 of a flowing liquid has some substantial advantages and therefore it is examined in the next sections of the present chapter.

2.9. Method of Varying Demagnetizing Volume [46]

The method is based on the relation of the amplitude of the nuclear resonance signal in a flow detector to the parameters of the polarizing apparatus. The liquid is polarized by flowing through a volume v_p in a strong magnetic field and then it flows through a connecting tube of volume v_t into a nuclear resonance detector.

The relation of the signal amplitude to the parameters of the polarizing apparatus has the form

$$A = k\left(1 - e^{-\frac{v_p}{qT_1}}\right) e^{-\frac{v_t}{qT_1}}, \qquad (1.9)$$

where q is the liquid flow rate and k is a coefficient independent of v_t and v_p.

By measuring the signal amplitude at two values of v_t or v_p with q constant it is possible to determine the relaxation time. For example, if a signal amplitude A_1 is observed with a connecting tube volume v_t and a signal amplitude A_2 with a volume $v_t + \Delta v$, from expression (1.9) the ratio of these amplitudes

$$\frac{A_1}{A_2} = e^{\frac{\Delta v}{qT_1}}, \qquad (2.9)$$

whence it is possible to find T_1:

$$T_1 = \frac{\Delta v}{q \ln \frac{A_1}{A_2}}. \qquad (3.9)$$

Formula (3.9) is valid if the liquid passes through the volume Δv with a velocity which is uniformly distributed across the section. This may be achieved by suitable design of the demagnetizing volume, using as a control the method described in Section 1.10.

In principle, it is also possible to use for the measurement a change in v_p, but in this case, in addition to a uniform distribution of the velocity across the section, it is necessary to guarantee high uniformity of the field within this volume. In practice, it is simpler to measure T_1 with a change in v_t and therefore theoretical calculations of the apparatus parameters and errors are given for this method.

3.9. Optimal Value of Δv and Minimal Measurement Error

The relative error of the measurements from formula (3.9)

$$\sigma T_1 = \frac{\Delta (\Delta v)}{\Delta v} + \frac{\Delta q}{q} + \frac{\frac{\Delta A_1}{A_1} + \frac{\Delta A_2}{A_2}}{\ln \frac{A_1}{A_2}}, \qquad (4.9)$$

where Δq is the indeterminacy of the set flow rate, $\Delta(\Delta v)$ is the error in the measurement of the set variable volume, and ΔA is the error in the measurement of the signal amplitude.

In practice, $\Delta A_1 = \Delta A_2 \approx A_{no}$, where A_{no} is the amplitude of the detector noise. By substituting these values in relation (4.9) we obtain an expression for σT_1 in terms of the signal amplitude

$$\sigma T_1 = \frac{\Delta (\Delta v)}{\Delta v} + \frac{\Delta q}{q} + \frac{1 + \frac{A_1}{A_2}}{a \ln \frac{A_1}{A_2}}, \qquad (5.9)$$

where $a = A_1/A_{no}$ is the signal-to-noise ratio when $\Delta v = 0$.

By substituting A_1/A_2 from formula (2.9) we obtain an expression for σT_1 in terms of Δv, q, and T_1:

$$\sigma T_1 = \frac{\Delta (\Delta v)}{\Delta v} + \frac{\Delta q}{q} + \frac{(1 + e^{\frac{\Delta v}{qT_1}})qT_1}{a \Delta v}. \qquad (6.9)$$

By ensuring that the measurement of the volume and flow rate is sufficiently accurate, it may be arranged that in expression (6.9) the first two terms of the right-hand part are much smaller than the third term and may be neglected. In this case the measurement error

$$\sigma_{T_1} = \frac{(1 + e^{\frac{\Delta v}{qT_1}}) \, qT_1}{a \Delta v} \,. \tag{7.9}$$

Let us find the value of Δv_{opt} with which this error is minimal. It may be determined from the condition

$$\frac{\partial (\sigma_{T_1})}{\partial (\Delta v)}_{\Delta v = \Delta v_{opt}} = 0. \tag{8.9}$$

By substituting the condition (8.9) σ_{T_1} from expression (7.9), we obtain after simplification

$$e^{\frac{\Delta v_{opt}}{qT_1}} \left(\frac{\Delta v_{opt}}{qT_1} - 1 \right) \approx 1. \tag{9.9}$$

Equation (9.9) is satisfied when

$$\Delta v_{opt} = 1.28 \, qT_1. \tag{10.9}$$

By substituting $\Delta v = \Delta v_{opt}$ from condition (10.9) in formula (2.9), we obtain

$$\frac{A_1}{A_2} = 3.6, \tag{11.9}$$

and by substituting the same value $\Delta v = \Delta v_{opt}$ in formula (7.9) we obtain the minimal measuring error

$$\sigma_{T_1 min} \approx \frac{3.6}{a} \,. \tag{12.9}$$

4.9. Range of Measurement of T_1

The measurement of very long and very short relaxation times is difficult because of the low signal-to-noise ratio, which is determined by the expression

$$a = a_0 (1 - e^{-\frac{v_p}{qT_1}}) \, e^{-\frac{v_t}{qT_1}}, \tag{13.9}$$

where a_0 is the signal-to-noise ratio under optimal conditions of magnetization when $v_p \gg qT_1$ and $v_T \ll qT_1$.

At the maximum relaxation time $T_{1 \, max}$, the signal-to-noise ratio is

$$a_{(T_1 = T_{1 \, max})} = a_0 \left(1 - e^{-\frac{v_p}{qT_{1 \, max}}} \right) . \tag{14.9}$$

From expression (12.9) the measurement error is given by:

$$\sigma_{T_1} = \frac{3.6}{a_0 (1 - e^{-\frac{v_p}{qT_{1 \, max}}})} \,. \tag{15.9}$$

The upper limit of the range of measurement may be determined from expression (15.9) by using the condition $\sigma_{T_1} < \sigma T_1 \text{per}$ ($\sigma T_1 \text{ per}$ is the maximum permissible error in the measurement of T_1),

$$\frac{3.6}{a_0\left(1-e^{-\dfrac{v_p}{qT_1 \max}}\right)} < \sigma_{T_1}\text{per} . \tag{16.9}$$

After simple rearrangement, formula (16.9) gives an expression for the upper limit of the range of measurement

$$T_{1\max} \leqslant \frac{v_p}{q \ln \dfrac{a_0 \sigma_{T_1}\text{per}}{a_0 \sigma_{T_1}\text{ per}-3.6}} . \tag{17.9}$$

At the minimum relaxation time $T_1 \min$, from expression (13.9) we obtain

$$a_{(T_1 = T_1 \min)} = a_0 e^{-\dfrac{v_t}{qT_1 \min}} \tag{18.9}$$

By substituting the value of a from expression (18.9) in formula (12.9) we obtain the error in the measurements at the lower limit of the measurement range

$$\sigma_{T_1} = \frac{3.6}{a_0 e^{-\dfrac{v_t}{qT_1 \min}}} . \tag{19.9}$$

By using the condition $\sigma_{T_1} < \sigma T_1 \text{per}$, from this formula we find an expression for the lower limit of the measurement range

$$T_{1\min} \geqslant \frac{v_t}{q \ln \left(\dfrac{a_0 \sigma_{T_1}\text{per}}{3.6}\right)} . \tag{20.9}$$

5.9. Parameter Requirements of Instrument for Measuring T_1

Requirements for Strength of Field H_a in Absorption Detector and Its Volume v_a.

If the optimal conditions $v_a/q \ll T_{2n} \ll T_{1n}$ and $\gamma H_1 v_a/q = 3\pi/4$, are fulfilled, the relation of the signal amplitude to the magnetization of the nuclei entering the detector M has the form:

$$A = kM \left[1 - \frac{X_0 H_a Z T_{1n}}{MT_1} + 0.4 \frac{X_0 H_a v_a}{MT_1 q} \right], \tag{21.9}$$

where k is a factor independent of M and Z is the saturation factor, given by

$$Z = \frac{1}{1+\gamma^2 H_1^2 T_{1n} T_{2n}} \approx \frac{1}{\gamma^2 H_1^2 T_{1n} T_{2n}} \approx \frac{v_a^2}{5.5 T_{1n} T_{2n} q^2} .$$

Let us denote the expression in brackets by [1 + B(M)], where

$$B(M) = 0.2 \frac{X_0 H_a}{M} \cdot \frac{v_a}{qT_1} \left(2 - \frac{v_a}{qT_{2n}}\right).$$ (22.9)

Then the signal amplitude

$$A = kM[1 + B(M)].$$ (23.9)

For the basic formula (3.9) it is assumed that there is exact proportionality between A and M, otherwise it will have the form

$$T_1 = \frac{\Delta v}{q \ln \dfrac{M_1}{M_2}},$$ (24.9)

where M_1 is the magnetization of the nuclei entering the working volume of the detector when $\Delta v = 0$ and M_2 is the magnetization when $\Delta v \neq 0$.

Let us substitute in formula (24.9) the value of M found in terms of A [expression (23.9)]:

$$T_1 = \frac{\Delta v}{q \ln \dfrac{A_1(1+B_2)}{A_2(1+B_1)}},$$ (25.9)

where A_1 and B_1 are the values of A and B when $M = M_1$ and A_2 and B_2 are the values of A and B when $M = M_2$.

Expression (25.9) is rearranged:

$$T_1 = \frac{\Delta v}{q \ln \dfrac{A_1}{A_2}\left[1 + \dfrac{\ln \dfrac{1+B_2}{1+B_1}}{\ln \dfrac{A_1}{A_2}}\right]}.$$ (26.9)

As $B \ll 1$, then

$$\ln \frac{1+B_2}{1+B_1} = \ln(1+B_2) - \ln(1+B_1) = B_2 - B_1.$$

In this case expression (26.9) has the form

$$T_1 = \frac{\Delta v}{q \ln \dfrac{A_1}{A_2}}\left(1 - \frac{B_2 - B_1}{\ln \dfrac{A_1}{A_2}}\right).$$ (27.9)

The additional error introduced into the measurement of T_1 by the term missing from formula (3.9) is given by

$$\sigma_{T_1} = \frac{|B_2 - B_1|}{\ln \dfrac{A_1}{A_2}}.$$ (28.9)

This error should not exceed the permissible error $\sigma_{T_1 per}$, and therefore the following condition should be fulfilled:

$$\frac{|B_2 - B_1|}{\ln \dfrac{A_1}{A_2}} \ll \sigma_{T_1 \text{per}} . \tag{29.9}$$

When $\Delta v = \Delta v_{\text{opt}}$, $\ln A_1/A_2 = 1.28$, and condition (29.9) has the form

$$|B_2 - B_1| \ll 1.3\sigma_{T_1 \text{per}} . \tag{30.9}$$

By substituting in this expression B_1 and B_2 from formula (22.9) we obtain

$$\frac{0.2 X_0 H_a v_a}{q T_1} \left(2 - \frac{v_a}{q T_{2n}} \right) \left| \frac{1}{M_2} - \frac{1}{M_1} \right| \ll 1.3\sigma_{T_1 \text{per}} . \tag{31.9}$$

If $\Delta v = \Delta v_{\text{opt}}$, then

$$\frac{1}{M_2} - \frac{1}{M_1} = \frac{2.6}{M_1} ,$$

and then formula (31.9) has the form

$$0.4 \frac{X_0 H_a}{M_1} \cdot \frac{v_a}{q T_1} \left(2 - \frac{v_a}{q T_{2n}} \right) \leqslant \sigma_{T_1 \text{per}}. \tag{32.9}$$

By substituting in formula (32.9) the value $M_{p_1} = X_0 H_n \exp(-v_t / q T_1)$, we obtain

$$\frac{H_a}{H_p} \cdot \frac{v_a}{q T_1} \left(2 - \frac{v_a}{q T_{2n}} \right) e^{\frac{v_t}{q T_1}} \leqslant 2.5\sigma_{T_1 \text{per}}. \tag{33.9}$$

As $v_a/q \ll T_{2n}$, expression (33.9) may be simplified:

$$\frac{H_a}{H_p} \cdot \frac{v_a}{q T_1} \leqslant 1.25\sigma_{T_1 \text{per}} e^{-\frac{v_t}{q T_1}} . \tag{34.9}$$

By substituting in the last formula the value $v_a = \pi d_a^3/4$ (d_a is the linear size of the detector), we obtain

$$\frac{H_a}{H_p} \leqslant \frac{1.6\sigma_{T_1 \text{per}} q T_{1 \min} e^{-\frac{v_t}{q T_{1 \min}}}}{d_a^3} . \tag{35.9}$$

Requirements for Strength of Stray Field H_t in the Region of the Demagnetiz-
ing Volume. By using the results presented in Ch. 1, we can write an expression for the magnetization of
the nuclei entering the working volume of the absorption detector,

$$M = X_0 H_p (1 - e^{-\frac{v_p}{q T_1}}) e^{-\frac{v_t + \Delta v}{q T_1}} \left[1 + \frac{H_t}{H_p} \frac{e^{\frac{v_t + \Delta v}{q T_1}} - 1}{1 - e^{-\frac{v_p}{q T_1}}} \right] . \tag{36.9}$$

125

From this expression it follows that

$$\frac{M_2}{M_1} = e^{\frac{\Delta v}{qT_1}} \frac{1 + \frac{H_t}{H_p} \dfrac{e^{\frac{v_t}{qT_1}} - 1}{1 - e^{-\frac{v_p}{qT_1}}}}{1 + \frac{H_t}{H_p} \dfrac{e^{\frac{v_t + \Delta v}{qT_1}} - 1}{1 - e^{-\frac{v_p}{qT_1}}}} = e^{\frac{\Delta v}{qT_1}} \frac{1 + D_1}{1 + D_2}, \tag{37.9}$$

where

$$D_1 = \frac{H_t}{H_p} \frac{(e^{\frac{v_t}{qT_1}} - 1)}{(1 - e^{-\frac{v_p}{qT_1}})}, $$

$$D_2 = \frac{H_t}{H_p} \frac{e^{\frac{v_t + \Delta v}{qT_1}} - 1}{1 - e^{-\frac{v_p}{qT_1}}}. \tag{38.9}$$

If the inequality (35.9) holds, then $M_1/M_2 = A_1/A_2$. By using formula (37.9) it is possible to write the relation of T_1 to A_1/A_2

$$T_1 = \frac{\Delta v}{q \ln \dfrac{A_1 (1 + D_1)}{A_2 (1 + D_2)}}. \tag{39.9}$$

Relation (39.9) is analogous to expression (25.9) and therefore we can immediately write a condition similar to condition (30.9),

$$|D_2 - D_1| < 1.3 \sigma_{T_1 \text{per}}. \tag{40.9}$$

This condition may be rearranged by using the values of D_1 and D_2 from relation (38.9),

$$\frac{H_t}{H_p} e^{\frac{v_T}{qT_1}} \left| \frac{e^{\frac{\Delta v}{qT_1}} - 1}{1 - e^{-\frac{v_p}{qT_1}}} \right| \leqslant 1.3 \sigma_{T_1 \text{per}}. \tag{41.9}$$

When $\Delta v = \Delta v_{\text{opt}}$,

$$\frac{H_t}{H_p} \leqslant \frac{\sigma_{T_1 \text{per}} (1 - e^{-\frac{v_p}{qT_1}}) e^{-\frac{v_t}{qT_1}}}{2}. \tag{42.9}$$

Relation Between the Volumes v_p and v_t and the Liquid Flow Rate. If we know the range of measureable values of T_1 and the signal-to-noise ratio given by the circuit, from expressions (17.9) and (20.9) we can obtain

126

$$\frac{v_p}{v_t} \gg \frac{T_{1\,\mathrm{max}}\ln\dfrac{a_0\sigma_{T_1\mathrm{per}}}{a_0\sigma_{T_1\mathrm{per}}-3.6}}{T_{1\,\mathrm{min}}\ln\dfrac{a_0\sigma_{T_1\mathrm{per}}}{3.6}}. \qquad (43.9)$$

In practice the volume v_p may be set by the dimensions of the available polarizing magnet and then the maximum value of v_t is determined from expression (43.9). If the diameter of the tube is set, then the volume v_t is determined from constructional considerations and the volume v_p is determined from relation (43.9).

The liquid flow rate q is determined from expression (17.9) or (20.9). If the relaxation time varies over a narrow range close to some value T_1, then the flow rate is best selected from the condition for obtaining the maximum signal amplitude [formula (19.5)]. With a given liquid flow rate q in the system, which is most realistic in the planning of an industrial control system, the values of v_p and v_t are found from expressions (17.9) and (20.9).

6.9. Example of Selection of Instrument Parameters

Let us set the range of relaxation times measured, $T_{1\,\mathrm{min}} = 0.5$ sec and $T_{1\,\mathrm{max}} = 2$ sec, the maximum measurement error, $\sigma_{T_1\mathrm{per}} = 0.05$, and the liquid flow rate in the system, 20 cc/sec. The signal-to-noise ratio with complete polarization of the liquid a_0 depends on the quality and degree of complexity of the electronic circuits and quite a realistic figure is $a_0 = 100$. From expressions (17.9 and (20.9) we find v_p and v_t: $v_p \geq 50$ cc and $v_t \leq 2.8$ cc. If we choose 0.3 cm as the diameter of the connecting tube, its length must not exceed 40 cm and with a diameter of 0.4 cm, the length must not exceed 26 cm. From expression (10.9), the optimal variable volume for $T_1 \approx 1$ sec equals 26 cc. The strength of the magnetic field in the NMR detector, according to expression (35.9), with a detector diameter less than 0.8 cm may equal the strength of the polarizing field and with a diameter of 10 cm, it should be less by a factor of two. From formula (42.9), the strength of the stray field H_T must be less by a factor of 80 than the strength of the polarizing field.

7.9. Effect of Nonuniformity of the Curve of Liquid Flow Rates Through a Section of the Measuring Volume on the Measurement Results

We arbitrarily divide the liquid flow through the measuring volume into n tubes of flow with cross sections ΔS_i and a length equal to the length of the measuring volume l. Within the i-th tube of flow the mean velocity of the liquid may be regarded as constant and equal to W_i. Liquid with a magnetization vector M_0 enters the measuring volume. The liquid passing through the i-th tube of flow remains within it for a time $t_i = l/W_i$ and its magnetization at the outlet of the measuring volume is $M_i = M_0 \exp(-l/W_i T_1)$. In unit time from this tube of flow there passes $\Delta S_i W_i$ cc of liquid, i.e., the i-th tube of flow transmits in unit time a magnetic moment of the nuclei equal to

$$m_i = \Delta S_i W_i M_0 e^{-\frac{l}{W_i T_1}}.$$

All n tubes of flow transmit in one second a magnetic moment

$$m = \sum_{i=1}^{n} m_i = M_0 \sum_{i=1}^{n} \Delta S_i W_i e^{-\frac{l}{W_i T_1}},$$

which is contained in $\displaystyle\sum_{i=1}^{n} \Delta S_i W_i = S_0 W_{av}$ cc of liquid (S_0 is the cross section of the measuring volume W_{av} the average velocity of the liquid).

Thus, the liquid emerging from the measuring volume has a magnetization

$$M = M_0 \frac{\sum\limits_{i=1}^{n} \Delta S_i W_i e^{-\frac{l}{W_i T_1}}}{S_0 W_{av}} . \tag{44.9}$$

Let $W_i = W_{av}(1 + \alpha_i)$, where $\alpha_i = (W_i - W_{av})/W_{av}$ is the relative deviation of the average velocity in the i-th tube from W_{av}.

Then expression (44.9) will have the form

$$\frac{M}{M_0} = \frac{1}{S_0 W_{av}} \sum_{i=1}^{n} \Delta S_i W_{av}(1 + \alpha_i) e^{-\frac{l}{T_1 W_{av}(1 + \alpha_i)}} . \tag{45.9}$$

When $(l/T_1 W_{av}) \alpha_i \ll 1$, neglecting terms of the third order of smallness and using the formulas $1/(x + 1) = 1 - x + x^2 \ldots$ and $e^x = 1 + x - x^2/2 \ldots$ when $x < 0.2$, it is possible to make the following rearrangement:

$$e^{-\frac{l}{W_{av} T_1 (1 + \alpha_i)}} \approx e^{-\frac{l}{W_{av} T_1}(1 - \alpha_i + \alpha_i^2 - \ldots)} \approx e^{-\frac{l}{W_{av} T_1}} \left\{ 1 + \frac{l}{W_{av} T_1} \left[\alpha_i + \alpha_i^2 \left(\frac{l}{2 W_{av} T_1} - 1 \right) \right] \right\}.$$

By substituting this value in expression (45.9) we obtain

$$\frac{M}{M_0} = \frac{e^{-\frac{l}{T_1 W_{av}}}}{S_0 W_{av}} \left[\sum_{i=1}^{n} \Delta S_i W_{av} + \sum_{i=1}^{n} \Delta S_i W_{av} \alpha_i \left(1 + \frac{l}{W_{av} T_1} \right) + \sum_{i=1}^{n} \Delta S_i W_{av} \alpha_i^2 \left(\frac{3l}{2 W_{av} T_1} - 1 \right) \right]. \tag{46.9}$$

Taking into account the obvious equalities $\sum\limits_{i=1}^{n} \Delta S_i = S_0$ and $\sum\limits_{i=1}^{n} \Delta S_i \alpha_i = 0$, from expression (46.9) we obtain

$$\frac{M}{M_0} = e^{-\frac{l}{T_1 W_{av}}} \left(1 + \frac{\sum\limits_{i=1}^{n} \Delta S_i \alpha_i^2 \frac{l^2}{2 T_1^2 W_{av}^2}}{S_0} \right) . \tag{47.9}$$

From this relation it is possible to obtain the difference between the relaxation time determined with formula (3.9) and the true relaxation time T_1.

$$\frac{\Delta T_1}{T_1} = \frac{l}{2 W_{av} T_1} \bar{\alpha}^2, \tag{48.9}$$

where $\alpha^2 = \frac{1}{S_0} \sum\limits_{i=1}^{n} \Delta s_i \frac{(W_i - W_{av})}{W_{av}}$ is the relative mean square deviation of the velocity from W_{av}.

The treatment presented is valid with the following conditions:

1) the signal amplitude is proportional to the liquid flow rate, i.e., there is high saturation in the detector; 2) there is no transverse mixing of the liquid in the stream.

In this case the error introduced by the spread of velocities is maximal. In actual fact, a similar treatment in cases when the signal amplitude is independent of the liquid flow rate or there is vigorous mixing of the liquid across the stream gives the following expression for ΔT_1:

$$\frac{\Delta T_1}{T_1} = \left(\frac{l}{2 W_{av} T_1} - 1 \right) \bar{\alpha}^2. \tag{49.9}$$

With the optimal demagnetizing volume, from expression (48.9) we find that $\Delta T_1/T_1 = 0.64\,\bar{\alpha}^2$ and from expression (49.9), $\Delta T_1/T_1 = -0.36\bar{\alpha}^2$. Thus, to avoid an additional error in the measurement of T_1 because of the spread of molecular velocities in the liquid stream it is necessary to fulfill the condition:

$$\bar{\alpha}^2 < 1.5 \quad \sigma_{T_1} \text{per.} \tag{50.9}$$

The value of $\bar{\alpha}^2$ may be measured by the method of magnetic labeling of the nuclei described in Section 1.10. For this it is necessary to label the nuclei in the liquid in front of the measuring volume Δv, to determine the law of the change in the signal amplitude in the detector when the labeling front passes through it, and to construct the relation of dN/dW to W similar to the curve given in Fig. 3.10. The value of $\bar{\alpha}^2$ is determined from the expression

$$\bar{\alpha}^2 = \frac{\mu_2}{W_{av}^2}, \tag{51.9}$$

where μ_2 is the central moment of the second order of the curve of dN/dW.

In practice, with developed turbulent flow $\overline{\alpha^2} = 0.01 - 0.002$ i.e., in this case without any special measures, the spread of molecular velocities in the liquid introduces a sufficiently small additional error into the measurement of T_1.

With laminar flow, the simplified method of estimating the error, given in the present section, is unsuitable as in this case the condition $(l/T_1 W_{av})(W_i - W_{av}) \ll 1$ does not hold. For this purpose it is possible to use the relation of the signal amplitude to T_1 with laminar flow. In the simplest case, when the signal amplitude is independent of the liquid velocity in the detector, it is determined by the expression:

$$A = A_0 \left[e^{-\frac{l_0}{2 W_{av} T_1}} + \frac{l_0}{2 W_{av} T_1} E_i \left(-\frac{l_0}{2 W_{av} T_1} \right) \right]$$

and with the signal amplitude proportional to the liquid velocity in the detector

$$A = A_0 \left[e^{-\frac{l_0}{2 W_{av} T_1}} - \frac{l_0}{2 W_{av} T_1} e^{-\frac{l_0}{2 W_{av} T_1}} - \frac{l_0^2}{4 W_{av}^2 T_1^2} E_i \left(-\frac{l_0}{2 W_{av} T_1} \right) \right]$$

A_0 is the signal amplitude when $l_0/W_{av} T_1 \ll 1$. This relation differs considerably from the simple exponential relation which holds for turbulent flow. It is much more complex as it changes with a change in the strength of the resonance oscillating field and other parameters of the detector. Therefore, measuring T_1 with laminar flow of the liquid requires a special calibrated instrument.

8.9. Description of Experimental Apparatus for Measuring T_1 by the Method of Varying Demagnetizing Volume

The procedure with a change in the demagnetizing volume v_t was used in the experimental apparatus built for measuring the relaxation time in a flowing liquid. The liquid was polarized in an iron-clad magnet with an interpolar space v_p = 400 cc. The strength of the polarizing field was 10,000 oe. The diameter of the connecting tube was 0.4 cm, its length 100 cm, and the volume v_t = 12.6 cc. The strength of the stray field where the connecting tube and volume Δv lay was 0.5 oe. The nuclear resonance detector had a volume of 15 cc and was placed in a field with a strength of 30 oe. The liquid was pumped with a centrifugal pump. The flow rate was kept at 74 cc/sec. The signal was detected with an autodyne system and the signal-to-noise ratio when $v_t \ll qT_1$ was about 30. The range of measurement $T_{1\,max}$ = 6 sec and $T_{1\,min}$ = 0.3 sec could be obtained with σT_{1per} = 5% from formulas (17.9) and (20.9).

Two variations on the instrument design were used. In one Δv was changed smoothly by means of a calibrated variable volume. In the second there were several calibrated volumes, each of which could be connected into the system. In the first case the measurements were carried out in the following way. With the minimum volume (Δv = 0) the signal amplitude was determined (A_1) and then by an increase in the variable volume, the signal amplitude was adjusted to A_2 = $A_1/3.6$. The value of the variable volume $\Delta v_{t\,opt}$ established in this way was determined from a special scale and the relaxation time required could be found from the formula T_1 = $\Delta V_{t\,opt}/1.28\,q$. This method is good in that it always guarantees the optimal measurement conditions, i.e., the minimum error σT_1 = 3.6/a.

In the second variant, the amplitude A_1 was determined when Δv = 0, then after one of the calibrated volumes had been connected into the system (Δv) the amplitude A_2 was determined. The volume connected in was that which was closest to Δv_{opt}. The relaxation time was found from the usual formula $T_1 = \Delta v/q \ln(A_1/A_2)$. This method gives a higher error as it is not always possible to select a calibrated volume with $\Delta v \approx \Delta v_{opt}$, but on the other hand there is no need of a variable volume, which is quite complex to construct as it must have a constant hydraulic resistance and a uniform distribution of the liquid velocities across the cross section over the whole range of measurements. The second method has the advantage in measurements of T_1 which vary over a narrow range, for example, in the determination of the temperature dependence of T_1. In this case it is possible to use one adjusted calibrated volume close to the optimal in size. One of the practical uses of this apparatus is described in the next section.

9.9. Application of Method of Measuring T_1

Recording Free Radicals and Triplet States Arising in Photolysis. The classical method of recording free radicals, namely, electron paramagnetic resonance, has a limited range of application. Its sensitivity is high only when the signal from radicals with a narrow electron resonance line is observed in a strong magnetic field.

Radicals with a broad line and even more so triplet states in solution may be observed much more efficiently through their effect on the spin-lattice relaxation time of the nuclei of the solvent. On this basis, A. I. Zhernovoi and S. P. Pivovarov developed a method of recording free radicals and triplet states formed in liquids during their irradiation. A flow detector with preliminary polarization was used for measuring the relaxation time of nuclei of the liquid.

When there was no irradiation of the liquid passing through the demagnetizing volume v, the magnetization of the nuclei fell with the natural spin-lattice relaxation time T_1. If during the time that it passed through the demagnetizing volume, the liquid was subjected to the action of radiation, which produced paramagnetic centers with a concentration C, the demagnetization of the nuclei proceeded with a shorter relaxation time T_1^*, which was related to T_1 by the equation $1/T_1^* = 1/T_1 + kC$, where k is a coefficient depending on the nature of the paramagnetic centers. By measuring the amplitude of the nuclear resonance signal A in the absence of radiation and A^* during irradiation it is possible to determine

$$C = \frac{q \ln \dfrac{A}{A^*}}{v\,k}\,.$$

In the experimental apparatus built, a permanent magnetic with a field strength of 3200 oe was used for polarization of the liquid and observation of the signal. The signal was detected with a high-stability bridge circuit, supplied by a radio-frequency generator, synchronized with the precession of the nuclei by means of a superregenerator. The hermetically sealed flow system with a volume of 400 cc, made of glass, had an attachment for outgassing the liquid by flushing with inert gases. The demagnetizing system, which was made of quartz glass, lay outside the magnet gap. A radiation source was placed close to it. Calibration of the apparatus by means of stable free radicals showed that its sensitivity was of the order of 10^{15} paramagnetic centers per cc. Free radicals formed by photolysis of acetone with ultraviolet light were observed with this apparatus.

Investigation of the Time Dependence of the Corrosion Rate of a Metal in a Liquid [44]. One of the signs of the corrosion of a metal in a liquid is the appearance in the liquid of the ions of this metal. If the metal gives paramagnetic ions in the liquid, their appearance affects the relaxation time of the liquid. The change in the relaxation time of a 0.15% solution of HCl in water when the solution contains iron objects protected with an acid-resistant lacquer of poor quality is shown in Fig. 1.9. The fall in T_1 is related to the increase in the concentration of iron ions in the water. For direct determination of the concentration of iron ions, a calibration curve of T_1 against the concentration of $FeCl_3$ in water was plotted on the same apparatus and is shown in Fig. 2.9. Comparison of Figs. 1.9 and 2.9 made it possible to construct Fig. 3.9, which describes the increase in the concentration of iron ions during the corrosion of the metal in a weak acid solution. The slope of the curve at each point gives the differential corrosion rate.

Observation of Negative Hydration. One of the most important advantages of the method described for measuring the relaxation time is the fact that it makes it possible to measure T_1 of individual lines of the nuclear resonance spectrum. For this purpose it is sufficient to pass the liquid from the demagnetizing volume through the detector of an NMR spectrometer and observe the relation of the amplitude of the different lines to $\Delta \nu$. The relaxation time T_1 of protons of water and hypophosphite ions in solutions of KH_2PO_2 and NaH_2PO_2 at various concentrations and temperatures. The results for 12°C are given in Fig. 4.9. Curves 1 and 2 show the relation of T_1 of protons of water and hypophosphite to the concentration of the KH_2PO_2 solution and curves 3 and 4 show the relation of T_1 of protons of water and hypophosphite to the concentration of the NaH_2PO_2 solution. Figure 4.9 shows that in a dilute solution of potassium hypophosphite the relaxation time of the water is greater than the relaxation time of the hypophosphite protons, while in the sodium hypophosphite solution, the relaxation time of water protons is lower. In saturated solutions the relaxation times of water and hypophosphite protons are equal.

According to the theory developed by O. Ya. Samoilov [183], the hydration of ions in solutions shows as an effect on the translational motion of the water molecules closest to the ion. Ions with positive hydration (Na^+) reduce the mobility of the nearest water molecules, while water molecules near to the ions with negative hydration (K^+) become more mobile.

The relaxation time T_1 of water protons is directly related to the mobility of the molecules. With an increase in mobility, the relaxation time increases and with a decrease in mobility, the relaxation time of the water "near" to them is greater than the relaxation time of "distant" water and vice versa in the presence of sodium ions. In Fig. 4.9, curves 1 and 3 show the mean relaxation time of "near" and "distant" water. Curves 2 and 4 show the relaxation time of protons of hypophosphite ions, which is close to the relaxation time of water "distant" from K^+ and Na^+, as confirmed by the fact that the times T_1 of water and hypophosphite protons in saturated solutions are equal. Therefore, in the presence of potassium ions the mean relaxation time of "near" and "distant" water is greater than that of "distant" water (curve 1 lies above curve 2), while in the presence of sodium ions it is less (curve 3 lies below curve 4). In addition to the hydration effects, the viscosity of the solution affects the relaxation time so that T_1 falls with an increase in the concentration of the KH_2PO_2 solution. Studying the hydration through the difference in the

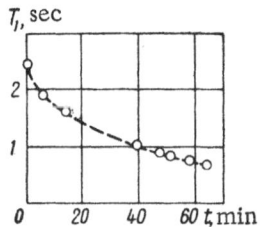

Fig. 1.9. Change in the relaxation time of a 0.15% solution of HCl in water as a result of the corrosion of iron.

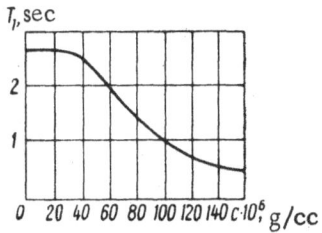

Fig. 2.9. Experimental relation of the relaxation time of $FeCl_3$ solution in water to the concentration.

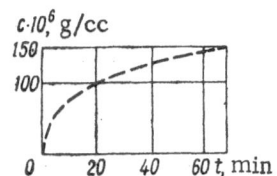

Fig. 3.9. Change in concentration of iron ions with time in the corrosion of iron in a 0.15% aqueous solution.

relaxation times of water and hypophosphite protons makes it possible to allow for this effect. The fact that the relaxation times T_1 of water and hypophosphite protons in saturated solutions are equal and their dependence on the presence of K^+ and Na^+ shows that hydration effects do not appear in this case.

10.9. Method of Measuring T_1 by Using Nutation of Magnetization of Nuclei

The drawback of the method of measuring T_1 examined in the previous sections is the need for changing the demagnetization volume, which makes automation of the measurements difficult. A method which does not have this drawback is described below.

A block diagram of the apparatus is shown in Fig. 5.9. The liquid is polarized by passing through a volume v_p lying in the field H of a permanent magnet. The volume must satisfy the condition for total polarization:

$$v_p \gg q_{max} T_{1max}$$

where q_{max} is the maximum liquid flow rate in the system and $T_{1\,max}$ is the maximum relaxation time of the liquid.

The liquid flowing from the volume v_p has a magnetization of the nuclei $M_p \approx X_0 H$, where X_0 is the static nuclear magnetic susceptibility. The liquid flowing from the volume v_p passes through the section of the tube lying inside the radiofrequency coil 1 (nutation coil), whose role will be explained later, then passes through the volume v_t and enters the coil of the nuclear magnetic resonance detector 2, where it gives a signal.

In contrast to the previous method, where the volume v_t must lie in a weak magnetic field, in this method the volumes v_p and v_t must be placed in fields of the same strength. For convenience, the device including coils 1 and 2 may be placed in the interpolar space of one magnet. When the liquid passes through the volume v_t the vector of the magnetization of the nuclei M changes according to the equation

$$M = M_p e^{-\frac{t}{T_1}} + X_0 H (1 - e^{-\frac{t}{T_1}}). \qquad (52.9)$$

As $M_p = X_0 H$, expression (52.9) assumes the form

$$M = X_0 H,$$

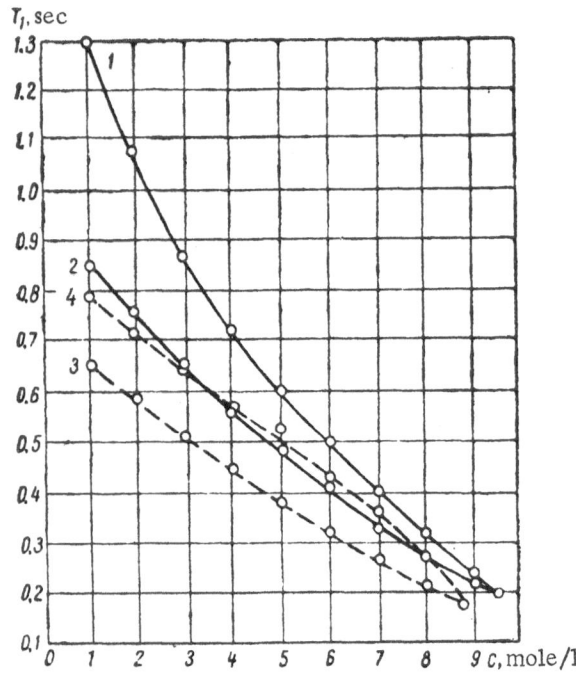

Fig. 4.9. Relation of the relaxation time T_1 of protons of water (curves 2 and 4) and hydrophosphite ions (curves 1 and 3) to the concentration of aqueous solutions of KH_2PO_2 and NaH_2PO_2: 1, 2) in KH_2PO_2 solution; 3, 4) in NaH_2PO_2 solution

i.e., the magnetization of the nuclei does not change when the liquid flows through the volume v_t and liquid with equilibrium magnetization flows into the coil of the nuclear resonance detector. Let us find what happens to M if a resonance oscillating field is excited in the nutation coil 1 to produce reorien-

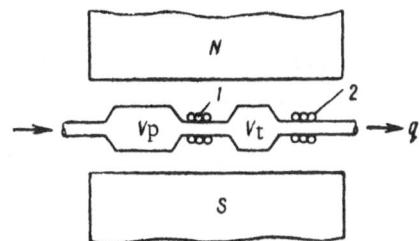

Fig. 5.9. Block diagram of the apparatus for measuring T_1: 1) nutation coil; 2) coil of nuclear resonance detector.

tation of the magnetization M_p. Into the volume v_t will flow liquid with magnetization FM_p, where F is the nutation coefficient, and expression (52.9) assumes the form

$$M' = FM_p e^{-\frac{t}{T_1}} + X_0 H (1 - e^{-\frac{t}{T_1}}). \qquad (53.9)$$

By substituting in the expression $t = v_t/q$, we obtain the magnetization of the nuclei in detector 2

$$M' = X_0 H [1 - (1 - F) e^{-\frac{v_t}{q T_1}}]. \qquad (54.9)$$

If the time for the liquid to flow through the coil of detector $2\tau \ll T_1$, polarization of the liquid will not occur in the detector and the signal amplitude will be proportional to the magnetization of the liquid flowing in.

In the absence of nutation the signal ampliutde A is proportional to M and when there is nutation, the signal amplitude A' is proportional to M'. The difference $A - A'$ is proportional to

$$M - M' = X_0 H (1 - F) e^{-\frac{v_t}{q T_1}}. \qquad (55.9)$$

The proportionality coefficient is independent of the magnetization and therefore we can write the expression

$$\frac{A - A'}{A} = (1 - F) e^{-\frac{v_t}{q T_1}}, \qquad (56.9)$$

whence

$$T_1 = \frac{v t}{q \ln \frac{A (1 - F)}{A - A'}}. \qquad (57.9)$$

The rotation factor F depends on the conditions in the nutation coil. If the magnetic field in the region of this coil is uniform, the most favorable condition, guaranteeing that F = −1, will be

$$\gamma H_1 \tau = \pi,$$

where H_1 is half the strength of the resonance oscillating field in the coil and τ is the time for the liquid to pass through the coil 1. This variant requires continuous fine adjustment of the field strength H_1 with a change in the value of τ due to variations in the liquid flow rate.

If the magnetic field is nonuniform across the liquid flow, by selecting a suitable value of H_1 it is possible to ensure that F = 0 independent of τ. With a gradient of the external field grad H, directed along the flow of the liquid, it is possible to guarantee that F = −(0.8-0.9), if condition (45.2) holds:

$$\frac{\gamma H_1^2}{W \, \text{grad} \, H} \geqslant 3,$$

where W is the velocity of the liquid in the cell 1. In this case F is independent of W over a wide range. For automation of the measurements, the resonance potential must be applied to the nutation coil periodically. In this case the signal amplitude will change periodically from A to A'. The maximum amplitude A may be measured after it has been rectified. The liquid flow rate q is determined from the lag of the moment that the signal changes behind the moment that the voltage on the nutation coil changes. If the lag is Δt, then $q \approx v_t/\Delta t$. The periodic application of the resonance potential to the nutation coil may be synchronized with the change in the signal: at the moment that the signal amplitude changes from A' to A the resonance potential must be applied to the nutaion coil and at the moment that the amplitude changes from A_0 to A', the resonance potential must be switched off. In this case the change in the signal and application of the potential to the coil will proceed with a frequency Ω, which is proportional to the liquid flow rate

$$\Omega = \frac{q}{2v_t}. \tag{58.9}$$

Thus, measurement of the maximum signal amplitude and the amplitude and frequency of the envelope signal makes it possible to obtain continuous information on the relaxation time and the liquid flow rate in the stream.

11.9. Optimal Parameters of Apparatus and Minimal Measurement Error

From expression (57.9) the relative error of the measurements is determined as

$$\sigma_{T_1} = \frac{\Delta v_t}{v_t} + \frac{\Delta q}{q} + \frac{\dfrac{\Delta A}{A} + \dfrac{\Delta A + \Delta A'}{A - A'} + \dfrac{\Delta F}{1 - F}}{\ln \dfrac{A(1-F)}{A - A'}}. \tag{59.9}$$

The first two terms represent the errors in the measurement of the volume v_t and the liquid flow rate q and they may be made sufficiently small so that the main contribution will be provided by the third term, which, after replacement of ΔA by $\Delta A'$, has the form

$$\sigma_{T_1} = \frac{\dfrac{A_{no}}{A} + \dfrac{2A_{no}}{A - A'} + \dfrac{\Delta F}{1 - F}}{\ln \dfrac{A(1-F)}{A - A'}} = \frac{\dfrac{A_{no}}{A}\left(1 + \dfrac{2}{(1-F)}e^{+\frac{v_t}{qT_1}}\right) + \dfrac{\Delta F}{1-F}}{\dfrac{v_t}{qT_1}}. \tag{60.9}$$

The nutation coefficient F with a given liquid flow rate and strength of the resonance oscillating field in the coil 1 (see Fig. 5.9) is constant and may be determined with a sufficiently high accuracy by repeated measurements. Let us assume that the error introduced by the indeterminacy of ΔF is negligibly small and then

$$\sigma_{T_1} = \frac{\left(1 + \dfrac{2}{(1-F)}e^{\frac{v_t}{qT_1}}\right)qT_1}{av_t}, \tag{61.9}$$

where a is the signal-to-noise ratio in the absence of nutation. In an investigation at the extremum of expression (61.9) the following condition is obtained for $v_{t\,opt}$:

$$\left(\frac{v_{t\,opt}}{qT_1} - 1\right)e^{\frac{v_{t\,opt}}{qT_1}} = \frac{1-F}{2}. \tag{62.9}$$

For the most favorable case $F = -1$, this condition does not differ from condition (9.9), i.e.,

$$v_{t\,opt} = 1.28qT_1, \tag{63.9}$$

when $F = 0$

$$v_{t\,opt} = 1.46qT_1.$$

By substituting the value of $v_{t\,opt}$ in expression (56.9) we obtain when $F = -1$

$$\frac{A_0 - A_0'}{A_0} = 2e^{-1.28} = 0.56,$$

whence

$$A_{0\,opt}' = 0.44A_0;$$

when $F = 0$

$$\frac{A_0 - A_0'}{A_0} = 0.28,$$

whence

$$A_{0\,opt}' = 0.72A_0.$$

By substituting $v_{t\,opt}$ in expression (61.9), we obtain when $F = -1$

$$\sigma_{T_1 min} \approx \frac{3.6}{a}\; ; \tag{64.9}$$

when $F = 0$

$$\sigma_{T_1 min} = \frac{6.6}{a}\; .$$

With a finite polarization volume v_p, in the liquid flowing from it the magnetization of the nuclei $M_p = X_0 H$ $[1 - \exp(-v_p/qT_1)]$. By substituting this value of M_p in expressions (52.9) and (53.9), after some rearrangements allowing for the fact that $\exp(-v_p/qT_1) \ll 1$, we obtain

$$\frac{M-M'}{M} = (1-F)e^{-\frac{v_t}{qT_1}}\left[1 - e^{-\frac{v_p}{qT_1}} + e^{-\frac{v_p+v_t}{qT_1}} - e^{-\frac{2v_p+v_t}{qT_1}}\right].$$

Let us denote the expression in square brackets by $(1-b)$, then

$$T_1 = \frac{v_t}{q\left(\ln\frac{M(1-F)}{M-M'} - b\right)}$$

or

$$T_1 = \frac{v_t}{q\ln\frac{A(1-F)}{A-A'}}\left(1 - \frac{b}{\ln\frac{A(1-F)}{A-A'}}\right)\; .$$

From this expression the relative error in the determination of T_1 due to the finite value of v_p equals

$$\sigma_{T_1} = \frac{b}{\ln\dfrac{A(1-F)}{A-A'}}.$$

With the optimal parameters of the apparatus, from expression (63.9) the value in the denominator equals 1.28, i.e.,

$$\sigma_{T_1} = 0.78\,b. \tag{65.9}$$

With allowance for the fact that $\exp(-v_{t\,opt}/qT_1)$, the expression for b has the form

$$b = 0.72\,e^{-\frac{v_p}{qT_1}} + 0.28\,e^{-\frac{2v_p}{qT_1}}.$$

The second term is considerably less than the first and may be neglected and then expression (65.9) assumes the form:

$$\sigma_{T_1} = 0.56\,e^{-\frac{v_p}{qT_1}},$$

whence we obtain the condition for the polarization volume

$$v_p \geqslant qT_1 \ln\frac{0.56}{\sigma_{T_1\,per}}, \tag{66.9}$$

where $\sigma T_{1\,per}$ is the permissible relative error in the measurement of T_1. On the size of the polarizing volume depends the signal-to-noise ratio $a = a_0\,[1-\exp(-v_p/qT_1)]$, where a_0 is the signal-to-noise ratio when $v_p \gg qT_1$. By substituting this value in expression (63.9) we obtain when $F = -1$

$$\sigma_{T_1} = \frac{3.6}{a_0\,(1-e^{-\frac{v_p}{qT_1}})};$$

and when $F = 0$

$$\sigma_{T_1} = \frac{6.6}{a_0\,(1-e^{-\frac{v_p}{qT_1}})},$$

whence we may find the required value of the signal-to-noise ratio when $F = -1$

$$\left.\begin{aligned} a_0 &> \frac{3.6}{\sigma_{T_1\,per}\,(1-e^{-\frac{v_p}{qT_1}})}; \\[2em] a_0 &> \frac{6.6}{\sigma_{T_1\,per}\,(1-e^{-\frac{v_p}{qT_1}})}. \end{aligned}\right\} \tag{67.9}$$

and when $F = 0$

136

By substituting the minimal value of v_p from relation (66.9) in formula (67.9) we obtain when $F = -1$

and when $F = 0$

$$
\left.
\begin{aligned}
a_0 &\geqslant \frac{3.6}{\sigma_{T_1 \text{per}} - 1.8\sigma_{T_1 \text{per}}^2} \; ; \\[2em]
a_0 &\geqslant \frac{6.6}{\sigma_{T_1 \text{per}} - 1.8\sigma_{T_1 \text{per}}^2} \; .
\end{aligned}
\right\}
\qquad (68.9)
$$

The method examined is most convenient to use for measuring T_1 which varies over a narrow range close to some definite value as in this case the parameters of the apparatus will be optimal. For example, by means of this apparatus it is possible to check the degree to which a liquid is freed from dissolved oxygen and other paramagnetic impurities.

The measurement error is determined by the stability of the liquid flow rate and the signal-to-noise ratio.

If the permissible measurement error is set, then it is possible to determine the lower limit of the required signal-to-noise ratio from expression (68.9). For example, when $\sigma_{T\text{per}} = 0.05$, $a_0 > 80$. The values of T_1 and q are determined by the parameters of the liquid and the flow system. The volume v_t is determined by the given values of q and T_1 from expression (63.9). The volume v_p is determined by the given values of q, T_1, and $\sigma_{T_1 \text{per}}$ from expression (66.9). For example, for pure water ($T_1 = 3$ sec) with $q = 10$ cc/sec and $\sigma_{T_1 \text{per}} = 0.05$, $v_t = 38.4$ cc and $v_p = 72.5$ cc, while for pure benzene ($T_1 = 19$ sec), the optimal values of v_t and v_p are as before if $q = 1.58$ cc/sec, while if $q = 10$ cc/sec, $v_t = 243$ cc and $v_p = 460$ cc.

CHAPTER 10

SOME OTHER APPLICATIONS OF NUCLEAR MAGNETIC RESONANCE

1.10 Investigation of Longitudinal Turbulent Diffusion in a Tube [44, 45]

Essence of Investigation Method. In the movement of a liquid in a tube with a constant cross section the mean velocity of the molecules equals the mean flow velocity of the liquid. The instantaneous velocity of an individual molecule differs from the mean velocity by a value called the pulsation velocity, which varies randomly both in magnitude and in direction. In a system of coordinates moving with the mean velocity of the liquid the motion of the molecules proceeds similarly to the normal molecular motion, but the velocities produced by the turbulent pulsations are added to the thermal velocities of the molecules. In connection with this, instead of the normal molecular diffusion, which occurs in a stationary liquid, there is much more effective turbulent diffusion of the molecules.

The direct method of investigating the diffusion of molecules through some section consists of labeling the molecules on one side of this section and determining after a definite time the number of molecules on the other side. In investigating diffusion the liquid is normally labeled by adding various dyes to it and the diffusion of the dye molecules is observed. This method is indirect as it is not the diffusion of the liquid molecules themselves which is observed, but that of the molecules of the additive. The diffusion of molecules may be observed directly by using radioactive labeling of the nuclei. Both of these methods are unsuitable for investigating turbulent diffusion in fast flows because of the difficulty of obtaining a sharp boundary of the labeled liquid.

Investigation of the nutation effect showed that by means of it it is possible to obtain in a fast stream of liquid a sharp boundary separating the polarized and unpolarized liquid and to observe its spread over a certain time. This is sufficient to investigate longitudinal turbulent diffusion.

To put this method into practice it is possible to use the normal apparatus for observing the nutation effect (see Fig. 2.2). It should contain a polarizing magnet with a strong field. After polarization in the interpolar space of the magnet the liquid enters a section of the tube of length l_0, in which the diffusion is investigated. At the beginning of this section a nutation detector coil is fixed onto the tube and at its end, an absorption detector coil, connected to a detector circuit with a short time constant.

When there is no resonance oscillating field in the nutation coil, a nuclear resonance signal is observed at the output of the detector circuit. At some moment of time a potential with a frequency equal to the precession frequency of the nuclei in the volume of the coil is applied to the nutation coil from an oscillator. Thereupon the magnetization of the nuclei in the coil at that moment falls rapidly to zero. This occurs in a time $T = 6/\gamma \Delta H_\perp$ where $\Delta H_\perp$ is the nonuniformity of the external field at right angles to the liquid flow in this coil. As a result of this depolarized liquid begins to flow from the coil, while through the tube there moves a boundary separating the polarized and depolarized liquid, which has a spread of $W_{av}T$ at the first moment. By producing a high transverse nonuniformity of the field in the coil, this spread may be made quite small. This boundary spreads out in passing through the tube as a result of turbulent diffusion. From the moment that the front edge of this boundary enters the absorption detector the nuclear resonance signal begins to fall and at the moment that its rear edge emerges from the detector the fall in the signal ceases. From the law of the change in the signal amplitude with time it is possible to find the law of the change in the magnetization of the nuclei through the diffusion layer.

Knowing the time t during which the total fall in the absorption signal amplitude occurred, it is possible to determine the length of the diffusion layer l_d and the turbulent diffusion coefficient is determined from this by the formula whose derivation is given below.

Let us write the basic diffusion equation

$$\frac{dP}{dt} = -D \frac{d\rho}{dx} S, \tag{1.10}$$

where P is the mass of substance diffusing, ρ is its density, x is the coordinate along which diffusion is being studied, S is the area of the surface through which diffusion occurs, and D is the diffusion coefficient.

In this case the role of ρ is played by the magnetization of the nuclei of the liquid M and instead of the mass P, the magnetic moment m is transferred as a result of diffusion. The coordinate x is directed against the flow of the liquid and the origin of the coordinates is best located in the section through which the boundary of the polarized and depolarized liquid passes at the initial moment and which moves with the mean velocity. In this case the equation for turbulent diffusion will have the form:

$$\frac{dm}{dt} = -D_t \frac{dM}{dx} S, \tag{2.10}$$

where D_t is the turbulent diffusion coefficient.

Practical investigation of the form of the depolarization front showed that when there is effective turbulent movement of the liquid the value dM/dx is independent of x. This follows, for example, from the linear relation of the signal amplitude to time shown in Fig. 3a.8.

In this case dM/dx is constant over practically the whole extent of the depolarization front and it may be divided by the length l_d:

$$\frac{dM(t)}{dx} = \frac{M_{max}}{l_d(t)} . \tag{3.10}$$

The time dependence of the length of the depolarization front may be represented by the expression

$$l_d = at + b. \tag{4.10}$$

When t = 0, $l_d = 6W_{av}/\gamma \Delta H_\perp = b$ the initial spread of the depolarization front; when t = $l_0/W_{av} = t_0$

$$l_d = a\frac{l_0}{W_{av}} + b, \text{ therefore } a = \frac{[l_d(t_0) - b] W_{av}}{l_0} .$$

By substituting formulas (3.10) and (4.10) in expression (2.10) we obtain

$$\frac{dm}{dt} = -D_t \frac{M_{max}}{at + b} . \tag{5.10}$$

Let us find the magnetic moment that has diffused after a time t_0,

$$\Delta m = \int_0^{t_0} \frac{dm}{dt} dt. \tag{6.10}$$

By substituting dm/dt from relation (5.10) in formula (6.10) and expanding a and b, we obtain

$$\Delta m = \frac{D_t M_{max} l_0}{\left[l_d(t_0) - \frac{6 W_{av}}{\gamma \Delta H_\perp} \right] W_{av}} \ln \left[\frac{l_d(t_0) \gamma \Delta H_\perp}{6 W_{av}} \right]. \tag{7.10}$$

The same value may be determined from the experimental curve in Fig. 3a.8. It equals the total magnetic moment of the nuclei behind the initial section.

$$\Delta m = \int_0^{\frac{l_d(t_0)}{2}} M(x) \, dx. \tag{8.10}$$

Figure 3a.8 shows that when x = 0

$$M(0) = \frac{M_{max}}{2},$$

and when $x = l_d(t_0)/2$

$$M\left(\frac{l_d(t_0)}{2} \right) = 0,$$

therefore

$$M_x = \frac{M_{max}}{l_d(t_0)} \left[\frac{l_d(t_0)}{2} - x \right].$$

By substituting this value in formula (8.10) and integrating we obtain

$$\Delta m = \frac{M_{max} l_d(t_0)}{8}. \tag{9.10}$$

By comparing expressions (7.10) and (9.10) we obtain the turbulent diffusion coefficient

$$D_t = \frac{l_d(t_0) W_{av} \left[l_d(t_0) - \frac{6 W_{av}}{\gamma \Delta H_\perp} \right]}{8 l_0 \ln \left[\frac{l_d(t_0) \gamma \Delta H_\perp}{6 W_{av}} \right]}. \tag{10.10}$$

In practice $6 W_{av}/\gamma \Delta H_\perp \ll l_d(t_0)$, and therefore, from formula (10.10) we obtain

$$D_t = \frac{l_d^2(t_0) W_{av}}{8 l_0 \ln \left[\frac{l_d(t_0) \gamma \Delta H_\perp}{6 W_{av}} \right]}. \tag{11.10}$$

Practical Measurements. A photograph of the resonance signal in a flow detector, changing as a result of a pulsed resonance oscillating field in the nutation coil, is shown in Fig. 1.10. The signal consists of a sinusoidal potential with a frequency of 550 Hz and the individual periods of this potential are not resolved on the photograph. The photograph was taken on motion picture film from the oscillograph screen. In the absence of diffusion the appearance and disappearance of the film should have occurred very sharply as the resonance frequency pulses ted to the nutation coil had a square form and therefore the sections of magnetized liquid in the stream had sharp front and rear edges when they were formed. The movement of the liquid in the stream

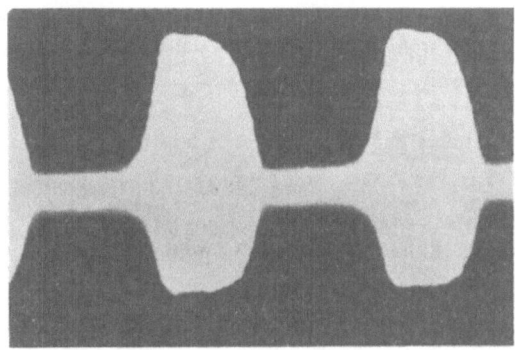

Fig. 1.10. Oscillogram of nuclear resonance signal with pulsed modulation of the magnetization of the stream.

produced the spread of the edges of the pulses of the nuclear resonance signal. In Fig. 1.10 the beginning of the pulse, which lies to the right and is produced by the arrival in the detector of the fastest molecules of the front edge of the magnetized liquid, is lost in noise, but then the end of the growth of the pulse, which indicates the arrival in the detector of the slowest molecules of the rear edge of the magnetized liquid, is clearly seen. The photograph clearly shows the beginning of the fall in the signal pulse produced by the arrival in the detector of the first portions of the demagnetized liquid, i.e., the fastest molecules of the rear edge of the magnetized liquid. The end of the fall in the pulse is lost in noise. The law for the fall in the signal pulse is identical with the law for its rise. This may be demonstrated by superposing one of the edges of the pulse on the mirror image of the other edge reflected in the time axis. The beginning of the front edge of the pulse, which is lost in noise, has a form analogous to the beginning of the rear edge. Thus, from the two edges of the pulse it is possible to establish completely the law of the change in nuclear resonance signal with time. To find the law of the change in the magnetization of the nuclei along the stream M(x) from the law of the change in the signal with time A(t) it is necessary to know the relation of the signal amplitude to the magnetization of the nuclei in the nuclear resonance detector. As was shown in Section 1.3, if the external field is sufficiently uniform, then the contribution to the signal amplitude from each elementary volume of the liquid, within which the velocity may be regarded as constant, is proportional to the magnetization of the nuclei of this volume and the value

$$\frac{1 - \cos \gamma H_1 \frac{l}{W}}{\gamma H_1 \frac{l}{W}}$$

where H_1 is half the strength of the resonance oscillating field in the detector, W is the velocity of the elementary volume averaged over the length of the detector l. If the field has considerable nonuniformity, then with a high strength H_1 the contribution to the signal amplitude from each elementary volume is proportional to the product of the magnetization of this volume ΔM and its velocity W, while the low value of H_1, the contributions to the signal amplitude are proportional to ΔM and independent of the velocity W.

It is quite obvious that the simplest relation between the signal amplitude and the magnetization of the nuclei will be found in precisely the latter case and therefore it is advantageous to select the parameters of the detector in apparatuses for studying the laws of hydrodynamics so that this relation is obtained. In this case the signal amplitude A is proportional to the mean magnetization and therefore it may be assumed that in relative units $\overline{M} = A$.

If the length of the detector is much less than the length of the depolarization front, the value of $\overline{M}$ in the detector equals the magnetization of the nuclei in the part of the stream which fills the detector at the given moment of time t i.e., is at a distance l_0 from the site of formation of the depolarization front. At the moment of time t, the detector contains only those molecules of polarized liquid which have not been able to pass through it and whose velocity $W \leq l_0/t$. The number of these molecules determines $\overline{M}$ and, consequently, the signal amplitude

$$A(W) \sim \overline{M}(W) \sim \int_{W_{min}}^{W = \frac{l_0}{t}} e^{-\frac{l_0}{W T_1} \frac{dN}{dW}} \, dW, \tag{12.10}$$

where (dN/dW) dW is the relative number of molecules in the stream whose velocity lies in the range from W to W + dW. This value is a continuous function of W and the signal amplitude at each moment of time t is uniquely related to the velocity $W = l_0/t$ of the molecules of the depolarization front passing through the detector at this moment.

Thus, from the relation of the amplitude of the nuclear resonance signal to time t it is possible to obtain the relation $\overline{M}(W)$ by replacing t by l_0/W.

Let us find the relation of the magnetization M to the coordinate x of the depolarization front at some fixed moment of time t_0. As was established, M is uniquely related to the velocity W of the molecules of the depolarization front which have the coordinates x at this moment, i.e., with a velocity

$$W = \frac{x}{t_0} + W_{av} = \frac{x + l_0}{t_0},$$

and therefore the relation M(x) may be obtained from the relation M(W) by substituting $W = (x + l_0)/t_0$ or directly from the relation M(t) by substituting $t = l_0 t_0/(x + l_0)$.

The relation of M to W obtained from the oscillogram in Fig. 1.10 is shown in Fig. 2.10. From the relaxation M(x) we can readily determine the length of the depolarization front l_d, which is required for determining the turbulent diffusion coefficient D_t. In practice, in a tube with a diameter of about 5 mm, with $W_{av} = 427$ cm/sec it was found that $D_t = 21$ cm^2/sec and with $W_{av} = 219$ cm/sec, $D_t = 5.3$ cm^2/sec, i.e., the turbulent diffusion coefficient increases rapidly with an increase in the liquid velocity.

The relation of dN/dW, obtained by differentiation of the curve in Fig. 2.10, is shown in Fig. 3.10. This value characterizes the distribution of the molecules with respect to velocities in the turbulent stream. If relaxation cannot be neglected then $dN/dW \sim (dM/dW) \exp(l_0/WT_1)$.

2.10. Measurement of Liquid Flow Rate from the Strength of the Resonance Oscillating Field in a Nutation Coil [194]

As was shown in Section 2.2, if a liquid passing along a tube is preliminarily polarized by a strong magnetic field and then passed successively through two nuclear resonance detectors, a resonance oscillating magnetic field in the coil of the first detector produces a change in the amplitude of the nuclear resonance signal detected in the second detector. Under certain conditions, in particular, if the strength of the external field in the first detector is less than the strength of the polarizing field, while its nonuniformity is low and directed along the flow of the liquid, then this change is proportional to $\gamma H_1 v_n/2q$, where H_1 is the strength of the resonance oscillating field in the coil of the first detector, γ is the gyromagnetic ratio of the nuclei, v_n is the volume occupied by flowing liquid in the coil of the first detector, and q is the liquid flow rate.

With an increase in H_1 the signal amplitude periodically becomes zero and the values H_{10} at which this occurs are strictly related to the liquid flow rate

$$H_{10} = \frac{q(2n-1)\pi}{\gamma v_n}, \tag{13.10}$$

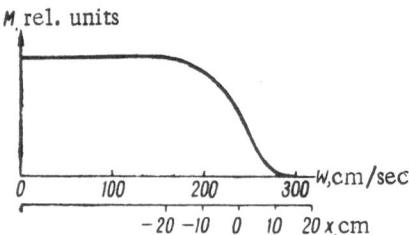

Fig. 2.10. Change in the magnetization of the nuclei across the spread of the depolarization front in a turbulent stream.

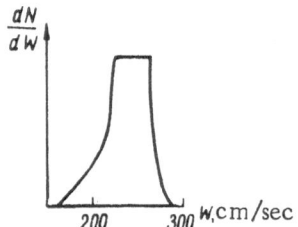

Fig. 3.10. Distribution of the molecules with respect to velocities in a turbulent stream.

where n is a whole number. Thus, knowing the strength of the resonance oscillating field in the coil of the first detector at which the amplitude of the nuclear resonance signal at the output of the detector circuit becomes zero it is possible to find the liquid flow rate

$$q = \frac{\gamma H_{10} v_n}{(2n-1)\,\pi} \, . \tag{14.10}$$

To determine the value of n, which must be substituted in expression (14.10), it is necessary to know how many times the signal amplitude becomes zero with an increase in H_1 from zero to H_{10}. If it is once then n = 1, if it is twice then n = 2, etc. In practice n may be more than 10.

Let us find the error in the measurement of the flow rate. It is assumed that it has been possible to establish the value $H_1 = H_{10}$ which exactly satisfies expression (13.10). In this case the signal amplitude A = 0 and the flow rate q may be found from expression (14.10). If the flow rate is now reduced, the increase in the signal amplitude is not detectable until it exceeds the noise level A_{no}. Thus, the change in the flow rate Δq at which the change in the signal amplitude $\Delta A = A_{no}$ is the absolute measurement error.

The change in the signal amplitude ΔA is related to a small change in the flow rate by the expression $\Delta A = (\partial A/\partial q)\,\Delta q$. By substituting $\Delta A = A_{no}$, we obtain

$$\Delta q = \frac{A_{no}}{\dfrac{\partial A}{\partial q}} \, . \tag{15.10}$$

The derivative $\partial A/\partial q$ consists of several terms, but only one of them does not contain the factor cos $\gamma H_1 v_n/2q$, which with $H_1 = H_{10}$, equals zero when $\gamma H_1 v_n/2q = (2n-1)\pi/2$ and therefore the other terms may be neglected. Then

$$\frac{\partial A}{\partial q} = A_0 \, \frac{\gamma H_1 v_n}{2q^2} \sin \frac{\gamma H_1 v_n}{2q} \, , \tag{16.10}$$

where A_0 is the amplitude of the nuclear resonance signal when $H_1 = 0$.

By substituting $\partial A/\partial q$ from formula (16.10) in expression (15.10), we obtain when $\gamma H_1 v_n/2q = (2n-1)\,\pi/2$

$$\frac{\Delta q}{q} = \frac{2}{a\,(2n-1)\,\pi} \, , \tag{17.10}$$

where a is the signal-to-noise ration when $H_1 = 0$.

In addition to the source of error considered, a contribution is also made by the inaccuracy of the measurement of the strength of the oscillating field H_1 and the volume v_n. The strength H_1 is proportional to the potential U on the coil of the first detector, which may be determined with high accuracy and therefore the relative error in the measurement of the liquid flow rate as a whole is determined by expression (17.10).

For absolute measurement of the liquid flow rate it is necessary to know the effective volume v_n and also the coefficient of proportionality between the potential U and the field strength H_1. These values may be measured reliably, while they are quite complicated to calculate and therefore it is advantageous to determine directly the coefficient of proportionality between the potential U and the liquid flow rate. The latter may be found independently with a sufficiently high accuracy. This method makes it possible to make accurate measurements of a flow rate with an error determined by the expression (17.10). The signal-to-noise ratio is maximal to some optimal flow rate. It falls with an increase or a decrease in the flow rate. The range of measurement is limited by the flow rates at which the signal-to-noise ratio falls to the value

$$a_{\min} = \frac{2}{\sigma_{q \text{ per}}(2n-1)\pi}, \qquad (18.10)$$

where $\sigma_{q \text{ per}}$ is the permissible relative error.

The method makes possible the automatic measurement and regulation of a flow rate as when a phase detector is used the sign of the nuclear resonance signal appearing depends on the sign of the mismatch between the liquid flow rate and the radiofrequency potential on the first detector coil.

The drawback of the method is the lag between the moment of the change in the signal amplitudes relative to the moment of a change in the flow rate, which is determined by the distance between the coils of the first and second detectors. This distance must be such that the frequency of the field H_1 is far from the frequency of the nuclear resonance detector.

Without any appreciable increase in the measurement error, the nutation detector may lie in a nonuniform field with the gradient directed along the liquid flow. This makes it possible to reduce the distance between the detectors.

The method may be used for the measurement and stabilization of the strength of the radiofrequency field H_1. For this purpose the movement of the liquid is better produced by a mechanical method such as a rotating disk rather than by its flow as in the former case it is much simpler to maintain a constant velocity and vary it over a wide range.

3.10. Measurement of Velocity of Liquid Dielectrics from the Instrument Frequency Shift [36]

If a section of the tube consists of a nuclear magnetic resonance detector, made in accordance with one of the designs shown in Fig. 1.4, the nuclear resonance effect in this detector is observed at a frequency of the oscillating field shifted from the nuclear resonance frequency by $\Delta\overline{\omega}_A$, which is proportional to the velocity of the liquid. The proportionality coefficient depends on the actual construction of the detector and is determined by expression (3.4). Knowing the geometric parameters of the detector, by measuring the frequency shift $\Delta\overline{\omega}_A$ it is possible to determine the absolute velocity and flow rate of the liquid.

For measuring $\Delta\overline{\omega}_A$ it is necessary to establish a certain magnetic field strength H in the volume of the detector, thus setting the nuclear precession frequency ω_0, and measure the resonance frequency of the oscillating field ω in the detector. A change in the field strength H by ΔH introduces into the measurement a relative error $\Delta q/q = \gamma \Delta H/\Delta\overline{\omega}_A$. For the error to be less than 1%, it must be guaranteed that $\Delta H < \Delta\overline{\omega}_A/\gamma$ 100. Usually, $\Delta\overline{\omega}_A/\gamma$ does not exceed a few hundredths of an oersted, hence we have the condition that the stability of the field must be of the order of 10^{-4} oe for a long period. This condition is difficult to fulfill. This may be avoided by connecting into the tube simultaneously two flow detectors with instrument shifts of opposite sign and measuring the difference in their resonance frequencies. In this case the measurement error is determined by the gradient of the field.

At the present time flowmeters have been developed on the basis of the instrument effect, using both the nuclear absorption signal in the field of a magnet and the signal of free precession in the earth's field.

4.10. Determination of the Sign of the Gyromagnetic Ratio of Nuclei

The sign of γ in experiments on nuclear resonance is determined by two methods. One of these is absolute [184, 185]. To achieve this it is necessary to detect the nuclear induction signal and for this the exciting field must be produced by a system of radiofrequency coils which guarantee rotation of the field in the direction of precession of the nuclear magnetic moments. Knowing the direction of rotation of the field and the direction of the external field it is possible to determine the sign of γ. Pickup from the rotating field in the receiver coil is balanced by a complex electronic system. The other method is relative [186, 187] and to use it it is necessary to obtain nuclear induction signals from the nuclei investigated and from nuclei with a known gyromagnetic ratio. The unknown sign of the gyromagnetic ratio is determined from the relative polarity of the

144

signals from these two types of nuclei. The method requires a change of the detector or the presence of two types of nuclei in the detector.

The method of determining the sign of γ in a flow detector [35] is based on the instrument effect. In observing the nuclear resonance signal from the nuclei investigated in Fig. 1.4, it may be seen that the sign of the instrument frequency shift depends on the sign of the gyromagnetic ratio of the nuclei. If the direction of the external field and the direction of rotation of the lines of force of the oscillating field acting on the moving nuclei are related by the right-hand screw rule, then the sign of the shift of the resonance frequency of oscillating field from the precession frequency of the nuclei in the detector is the same as the sign of γ. This method has the following advantages: 1) It makes it possible to determine any sign of γ in work with any type of nuclear resonance signal, i.e., with any detector circuit; 2) it is simpler than the known absolute method as it does not require a rotation field and compensation of the pickup from it; 3) it is simpler than the known relative method because it does not require comparison of signals from test and standard nuclei.

If the relaxation time of the liquid $T_1 > 0.1$ sec, then for determining the sign of γ there is no need to complicate the construction of the main detector, which could lead to deterioration of the signal. For this purpose it is possible to use the nutation effect in observing the shift in the resonance frequency in a toroidal detector, which is placed in a weak field and through which there flows the polarized liquid before entering the main detector, and to determine the presence of resonance from the change in the signal in the main detector.

5.10. Nuclear Magnetic Resonance Spectrometers with a Flowing Liquid

The first nuclear magnetic resonance spectra in a flow liquid were observed by Bloom and Shoolery [3]. They observed an increase in the signal amplitude as a result of the flow of the polarized liquid into the detector, but expressed doubts on the advantages of using such spectrometers for observing the hyperfine structure of spectra. In actual fact, the increase in the signal amplitude obtained as a result of the movement of the liquid resulted in broadening of the lines of the NMR spectrum. Nonetheless, nuclear magnetic resonance spectrometers with a flowing liquid have an important advantage. They make it possible to achieve remote control of the qualitative and quantitative composition of a liquid flowing continuously through a tube and this is very important in using these instruments in industry.

A flow spectrometer may be made from a normal spectrometer by replacing the sample ampoule in it by a tube through which preliminarily polarized liquid flows. The polarizer may be in the field of the same or another magnet and its volume must satisfy the condition that $v_p \gg qT_1$ (q is the liquid flow rate). The value of q and the volume of the detector must be selected so that the width of the line due to the nonuniformity of the field approximately equals the width of the line due to the liquid flow. These conditions were examined in Ch. 5. Considerable narrowing of the line may be obtained by using a flow detector with a rotating liquid [83]. For example, with the detector illustrated in Fig. 30.3 the line width of the NMR spectrum was about 0.001 oe. If there is no need for high resolution, for example, in isotopic analysis, to increase the signal amplitude it is advantageous to increase the cross section of the detector and the liquid flow rate. In this case it is possible to place the detector in a weak magnetic field to reduce the weight and increase the stability of the instrument without an appreciable loss in the signal amplitude.

In addition to the undoubted advantages of flow spectrometers in their industrial application, they make it possible to carry out some original experiments. For example, by means of a flow spectrometer with preliminary polarization built in France, Hennequin investigated the spectra of the hyperfine interaction of nuclei in a weak magnetic field, when the difference in the precession frequencies of the interacting nuclei was comparable with the constant hyperfine structure. On a flow spectrometer with a weak field fitted with a system for measuring the relaxation time by means of a variable demagnetizing volume (Ch. 9), it is possible to measure the relaxation time of individual lines of the NMR spectrum [17, 46]. On this apparatus the negative hydration of cations in solutions was observed for the first time by the nuclear resonance method and the experiments on double resonance described in the next section were also carried out.

6.10. Measurement of Spin Exchange Rate by the Double Nuclear Resonance Method

If a substance investigated contains nuclei which are not equivalent chemically, i.e., are in different chemical surroundings, nuclear resonance signals from these groups are observed at several different frequencies. It is possible to observe the resonance signal of the nuclei of each group and estimate their total magnetic moment from its amplitude. The presence of spin exchange between nuclei of separate groups may be observed from the line broadening of the nuclear resonance spectrum of these groups, which occurs as a result of the decrease in the lifetime of the individual states of the nuclei.

In using this effect for measuring the rate of spin exchange between the nuclei of different groups it is necessary to eliminate the contribution to the width of the nuclear resonance line introduced by spin exchange between nuclei of the same group (spin-spin relaxation), spin exchange of the nuclei with the lattice (spin-lattice relaxation), and the nonuniformity of the external field. This considerably reduces the accuracy of the measurements.

The presence of spin exchange between two groups of nuclei may be observed if the lifetime τ of the spin of the nucleus of one group with respect to exchange with the spin of the nucleus of another group has a value of the order of $1/\Delta f$, where Δf is the frequency difference which is observed between the resonance of the nuclei of the two groups. If changing some parameter such as the temperature, magnetic field, or concentration changes τ, then when $\tau \gg 1/\Delta f$ two nuclear resonance lines are observed, when $\tau \approx 1/\Delta f$, the two lines begin to merge, and when $\tau \ll 1/\Delta f$, the nuclei of the two groups give one nuclear resonance line.

This effect is used widely in practice for estimating τ, but it is not suitable for investigating slow exchange when $\tau \gg 1/\Delta f$.

Below we will describe a method of investigating spin exchange which makes it possible to measure τ over a wider range with higher accuracy than known methods. Let us examine the essence of the double resonance method on the example of the measurement of the rate of exchange of spins between n groups of nuclei, corresponding to n lines of the NMR spectrum. Let us denote the total magnetic moments of the nuclei in these states by $M_1, M_2, \ldots, M_n$ and the longitudinal relaxation times of the nuclei by $T_1^1, T_1^2, \ldots, T_1^n$. The change in $M_1, M_2, \ldots, M_n$ is described by the expressions:

$$\left.\begin{aligned}
\frac{dM_1}{dt} &= \frac{M_{01} - M_1}{T_1'} - \frac{M_1}{\tau_1} + \frac{M_2}{\tau_{21}} + \ldots + \frac{M_n}{\tau_{n_1}}, \\
\frac{dM_2}{dt} &= \frac{M_{02} - M_2}{T_1^2} - \frac{M_2}{\tau_2} + \frac{M_1}{\tau_{12}} + \ldots + \frac{M_n}{\tau_{n_2}}, \\
&\cdots\cdots\cdots\cdots\cdots\cdots\cdots\cdots\cdots\cdots\cdots\cdots \\
\frac{dM_n}{dt} &= \frac{M_{0n} - M_n}{T_1^n} - \frac{M_n}{\tau_n} + \frac{M_1}{\tau_{1n}} + \ldots + \frac{M_{n-1}}{\tau_{n-1,\,n}},
\end{aligned}\right\} \tag{19.10}$$

Where $M_{01}, M_{02}, \ldots, M_{0n}$ are the equilibrium values of the total magnetic moments of the nuclei in the states investigated, $\tau_1, \tau_2, \ldots, \tau_n$ are the lifetimes of the spins of the nuclei in the states $1, 2, \ldots, n$, respectively, with respect to exchange with all the other states, and τ_{ab} is the lifetime of the spins of the nuclei in state a with respect to exchange with spins of nuclei in state b.

If the substance remains for a sufficiently long time in a constant external magnetic field, then $M_1 = M_{01}$, $M_2 = M_{02}, \ldots, M_n = M_{0n}$, and $dM_1/dt = dM_2/dt = \ldots = dM_n/dt = 0$. In this case it follows from expression (19.10) that

$$\sum_{\substack{a=1 \\ a \neq b}}^{n} \frac{M_{0a}}{\tau_{ab}} = \frac{M_{0b}}{\tau_b}. \tag{20.10}$$

If at some moment of time t = 0, by the action of a resonance oscillating field line 1 is "saturated", i.e., M_1 is made equal to 0, leaving M_2, M_3,..., M_n at the equilibrium values, then as a result of exchange M begins to change according to a law, which may be written by substituting in expression (19.10) $M_1 = 0$, $M_2 = M_{02}$, $M_3 = M_{03}$ ... $M_n = M_{0n}$ and taking into account (20.10)

$$\left.\begin{aligned}
\frac{dM_1}{dt}_{t=0} &= M_{01}\left(\frac{1}{\tau_1} + \frac{1}{T_1'}\right), \\
\frac{dM_2}{dt}_{(t=0)} &= -\frac{M_{01}}{\tau_{12}}, \\
&\cdots\cdots\cdots\cdots\cdots\cdots \\
\frac{dM_n}{dt}_{(t=0)} &= -\frac{M_{01}}{\tau_{1n}}.
\end{aligned}\right\} \tag{21.10}$$

Thus, by saturating one of the lines of the NMR spectrum and measuring the rate of change of all the lines at the moment of time t = 0 it is possible to determine the relaxation time of the "saturated" line and the probability of its spin exchange with each line. The lifetime of the spins of the nuclei in the state corresponding to the "saturated" line τ_1 is determined by the sum of the probabilities of spin exchange with all the other nuclei

$$\frac{1}{\tau_1} = \sum_{a=2}^{n} \frac{1}{\tau_{1a}}. \tag{22.10}$$

Practical measurements of the rate of exchange of nuclear spins by the double resonance method are made on a flow NMR spectrometer with two detectors, the block diagram of which is given in Fig. 4.10. In this spectrometer the liquid investigated was polarized by flowing through the interpolar space of the magnet 1 with a field strength of 8 koe and then it flowed through the coil 2 of the first nuclear resonance detector, lying in the magnet 3. The liquid next flowed through a vessel of variable volume 4, lying in a weak magnetic field, and entered the second nuclear magnetic resonance detector 5, which lay in a uniform magnetic field with a strength of 35 oe, produced by a magnet 6. The coil of the second NMR detector was connected to the spectrometer circuit. In contrast to the normal method of observing an NMR signal in a flow detector with preliminary polarization, the system of nuclei emerging from the polarizer is not in equilibrium with the external magnetic field and has a magnetization much greater than the equilibrium value. The equilibrium magnetization in a field with a strength of 35 oe and other weak fields through which the liquid flows after the polarizer is close to zero. As was shown, for measuring the rate of exchange of nuclei between several lines it is necessary to ensure that all the lines except one are in equilibrium with the external field. Coil 2 is used for this purpose and it lies in a magnetic field with a strength of 0.14 oe with a nonuniformity within the coil no higher than 0.005 oe. By producing in this coil a resonance oscillating magnetic field of suitable strength and frequency it is possible to produce nutation of the nuclear magnetization of the required lines through an angle close to $\pi/2$. As the transverse relaxation time $T_2^* \ll T_1$ because of the nonuniformity of the external magnetic field, during the subsequent flow of the liquid the magnetization of these lines falls rapidly, i.e., it approaches the equilibrium value.

In practice, the "nutation" of the spectral line must be carried out in such a way that with the minimal value of the variable volume 4, in the liquid flowing into the second detector the magnetization of the corresponding groups of nuclei will equal zero and this may be checked as the amplitudes of the corresponding spectral lines should

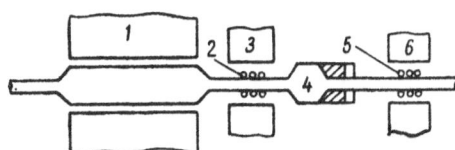

Fig. 4.10. Block diagram of flow spectrometer with two detectors: 1) polarizing magnet; 2) coil of first detector; 3) magnet of first detector; 4) variable volume; 5) coil of second detector; 6) magnet of second detector.

equal zero. If the variable volume is then increased by Δv, liquid will flow into the detector with values of M that do not equal zero, but are determined by exchange and relaxation processes.

Let us examine an example of measuring τ for the case of an aqueous solution of potassium hypophosphite, whose spectrum consists of three lines. The central line is produced by water protons and the side lines are produced by protons of hypophosphite ions $H_2PO_2^-$, which are separated from each other by 500 Hz because of indirect spin-spin interaction of the protons with the phosphorus nucleus.

In the case of three lines with amplitudes A_1, A_2, and A_3, which are proportional to the magnetizations M_1, M_2, and M_3, the law of their change may be found replacing M_1, M_2, and M_3 in expression (19.10) by A_1, A_2, and A_3 and substituting $t = \Delta v/q$ and $M_{01} = M_{02} = M_{03} = 0$ (q is the flow rate), when we obtain:

$$\left. \begin{array}{l} \dfrac{dA_1}{d\,\Delta v}\,q = -A_1\left(\dfrac{1}{T_1'} + \dfrac{1}{\tau_1}\right) + \dfrac{A_2}{\tau_{21}} + \dfrac{A_3}{\tau_{31}}\,, \\[3mm] \dfrac{dA_2}{d\,\Delta v}\,q = -A_2\left(\dfrac{1}{T_1^2} + \dfrac{1}{\tau_2}\right) + \dfrac{A_1}{\tau_{12}} + \dfrac{A_3}{\tau_{32}}\,, \\[3mm] \dfrac{dA_3}{d\Delta\,v}\,q = -A_3\left(\dfrac{1}{T_1^3} + \dfrac{1}{\tau_3}\right) + \dfrac{A_1}{\tau_{13}} + \dfrac{A_2}{\tau_{23}}\,. \end{array} \right\} \qquad (23.10)$$

As a result of nutation of the second and third lines, when $\Delta v = 0$:

$$A_1 = A_{10}, \quad A_2 = A_3 = 0.$$

By substituting these values in system (23.10) we obtain:

$$\left. \begin{array}{l} \dfrac{dA_1}{d\Delta v}_{(\Delta v=0)}\,q = -A_{10}\left(\dfrac{1}{T_1'} + \dfrac{1}{\tau_1}\right), \\[3mm] \dfrac{dA_2}{d\,\Delta v}_{(\Delta v=0)} = +\dfrac{A_{10}}{\tau_{12}}\,, \\[3mm] \dfrac{dA_3}{d\,\Delta v}_{(\Delta v=0)} = +\dfrac{A_{10}}{\tau_{13}}\,. \end{array} \right\} \qquad (24.10)$$

The lifetime of the spins of the nuclei τ_1 in the state corresponding to the first line is determined by the relation

$$\frac{1}{\tau_1} = \frac{1}{\tau_{12}} + \frac{1}{\tau_{13}}\,. \qquad (25.10)$$

As a result of nutation of the first line, when $\Delta v = 0$, $A_1 = 0$, $A_2 = A_{20}$, $A_3 = A_{30}$. By substituting these values in relation (23.10) we obtain:

$$\left. \begin{array}{l} \dfrac{dA_1}{d\,\Delta v}_{(\Delta v=0)}\,q = \dfrac{A_{20}}{\tau_{21}} + \dfrac{A_{30}}{\tau_{31}}\,, \\[3mm] \dfrac{dA_2}{d\,\Delta v}_{(\Delta v=0)}\,q = -A_{20}\left(\dfrac{1}{T_1^2} + \dfrac{1}{\tau_2}\right) + \dfrac{A_{30}}{\tau_{32}}\,, \\[3mm] \dfrac{dA_3}{d\,\Delta v}_{(\Delta v=0)}\,q = -A_{30}\left(\dfrac{1}{T_1^3} + \dfrac{1}{\tau_3}\right) + \dfrac{A_{20}}{\tau_{23}}\,. \end{array} \right\} \qquad (26.10)$$

Figure 5.10 gives the experimental relations of the total amplitude A_h of the two lines of the potassium hypophosphite protons to the variable volume Δv with a flow rate of 22.2 cc/sec. The amplitudes are given in scale divisions of the recording spectrometer.

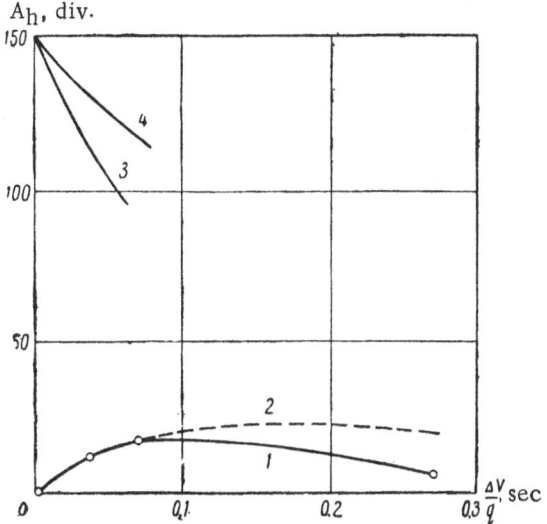

Fig. 5.10. Relation of the amplitude of hypophosphite protons in the second detector to the variable volume: 1, 2) line of hypophosphite protons "saturated" in first detector; 1) experimental curve; 2) theoretical curve; 3) line of water protons "saturated" in first detector; 4) no saturation.

Curve 1 was obtained with the amplitude and frequency of the oscillating magnetic field H_1 in the coil such that when $\Delta v = 0$, both the lines of the hypophosphite protons equaled zero. Curve 3 was obtained with H_1 such that when $\Delta v = 0$, the line of the water protons equaled zero. Curve 4 was obtained without conditions for nuclear resonance in coil 1. The case corresponding to curve 1 (see Fig. 5.10) is described by the system of equations (24.10) if A_1 is taken as the amplitude of the line of the water protons and A_2 and A_3 denote the amplitudes of the lines of the hypophosphite ion protons. This system shows that from the slope of curve 1 when $\Delta v = 0$ it is possible to determine the lifetime of the spins of the hypophosphite ion protons with respect to exchange with the spins of water protons:

$$\tau_h = \frac{A_{h0}}{\dfrac{dA_h}{d\,\Delta v}\, q}, \qquad (27.10)$$

where A_h is the total amplitude of the lines of the hypophosphite ion protons and A_{h0} is the value of A_h when $\Delta v = 0$.

The case corresponding to curve 3, with the same symbols is described by the system of equations (26.10), which shows that from the slope of curve 3 it is possible to determine the lifetime of the spins of the hypophosphite ion protons:

$$\frac{1}{T_{1h}} + \frac{1}{\tau_h} = \frac{dA_h}{d\,\Delta v}\, q / A_{h0}, \qquad (28.10)$$

where T_{1h} is the spin-lattice relaxation time of hypophosphite ion protons.

From curve 1 (see Fig. 5.10) it is possible to find the derivative $(dA_h/d\Delta v)_{\Delta v = 0}\, q$ = 114 divisions/sec. By substituting this value and A_{h0} = 150 divisions in expression (27.10) we obtain

$$\tau_h = 0.37 \text{ sec}$$

From curve 3 we find the derivative $(dA_h/d\Delta v)_{\Delta v = 0}$ = 114 divisions/sec. By substituting this value and A_{h0} = 150 divisions in expression (28.10) we obtain

$$\frac{1}{T_{1h}} + \frac{1}{\tau_h} = 7.6\, \frac{1}{\text{sec}},$$

whence $T_1 = 0.21$ sec.

This value coincides with the relaxation time of protons of water in the solution investigated, measured independently. As a check, Fig. 5.10 (curve 4) gives the relation of A_h to Δv without a saturating field in the coil 1. This case is described by the system of equations (23.10), which shows that if the relaxation times of water and hypophosphite ion protons coincide, from the slope of curve 4, independently of τ_h, it is possible to determine the relaxation time of hypophosphite ion protons:

$$T_{1h} = \frac{A_{h0}}{\dfrac{dA_h}{d\,\Delta v}\, q}. \qquad (29.10)$$

From curve 4, A_{h0} = 150 divisions and $(dA_h/d\Delta v)_{\Delta v = 0}$ = 550 divisions/sec and by substituting these values in expression (29.10) we obtain T_1 = 0.2 sec.

The accuracy of the values of τ_h and T_{1h} obtained may be checked by yet another method. The system of equations describing the change in the amplitude of the line of water protons A_w and the total amplitude of the lines of hypophosphite ion protons A_h has the following form:

$$\left.\begin{aligned} \frac{dA_w}{dt} &= -\left(\frac{1}{T_{1w}}+\frac{1}{T_w}\right)A_w+\frac{A_h}{\tau_h}, \\ \frac{dA_h}{dt} &= -\left(\frac{1}{T_{1h}}+\frac{1}{\tau_h}\right)A_h+\frac{A_w}{\tau_w}. \end{aligned}\right\}$$

(30.10)

By solving this system, we obtain the following relation of the amplitude A_h to Δv:

$$A_h = e^{-\frac{\Delta v}{2q}\left(\frac{1}{T_{1h}}+\frac{1}{T_{1w}}+\frac{1}{\tau_h}+\frac{1}{\tau_w}\right)}\left\{A_{h0}\,\mathrm{ch}\,\frac{\Delta v}{q}\sqrt{\left(\frac{1}{2T_{1h}}+\frac{1}{2\tau_h}-\frac{1}{2T_{1w}}-\frac{1}{2\tau_w}\right)^2+\frac{1}{\tau_h\tau_w}}\right.$$

$$+\frac{A_{h0}\left(\frac{1}{2T_{1h}}-\frac{1}{2T_{1w}}+\frac{1}{2\tau_h}-\frac{1}{2\tau_w}\right)+\frac{A_{w0}}{\tau_w}}{\sqrt{\left(\frac{1}{2T_{1h}}+\frac{1}{2\tau_h}-\frac{1}{2\tau_{1w}}-\frac{1}{2\tau_w}\right)^2+\frac{1}{\tau_h\tau_w}}}\,\mathrm{sh}\,\frac{\Delta v}{q}\left.\sqrt{\left(\frac{1}{2T_{1h}}+\frac{1}{2\tau_h}-\frac{1}{2T_w}-\frac{1}{2\tau_w}\right)^2+\frac{1}{\tau_h\tau_w}}\right\}.$$

(31.10)

On condition that $A_h = 0$, which is satisfied in the case of curve 1, expression (31.10) has the form:

$$A_h = e^{-\frac{\Delta v}{2q}\left(\frac{1}{T_{1h}}+\frac{1}{T_{1w}}+\frac{1}{\tau_h}+\frac{1}{\tau_w}\right)}\frac{A_{h0}}{\tau_h\sqrt{\left(\frac{1}{2T_{1h}}+\frac{1}{2\tau_h}-\frac{1}{T_{1w}}-\frac{1}{2\tau_w}\right)^2+\frac{1}{\tau_h\tau_w}}}$$

$$\times\,\mathrm{sh}\,\frac{\Delta v}{q}\sqrt{\left(\frac{1}{2T_{1h}}+\frac{1}{2\tau_h}-\frac{1}{2T_{1w}}-\frac{1}{2\tau_w}\right)^2+\frac{1}{\tau_h\tau_w}}.$$

(32.10)

The relation of A_h to $\Delta v/q$, obtained from the last expression after substituting $T_{1w} = T_{1h}$ = 0.21 sec, τ_h = 0.37 sec, and $A_{w0}/\tau_w = A_{h0}/\tau_h$ when A_{h0} = 150 divisions, is shown in curve 2 (see Fig. 5.10). Of the four experimental points of curve 1, three lie well on the theoretical curve 2 and only one point, which was obtained with $\Delta v/q$ = 0.27 sec, lies appreciably below. The reason for this discrepancy is apparently connected with the relation of τ_h to the strength of the magnetic field, which with the maximum volume Δv differs somewhat from the field at the two lower volumes.

Spin exchange between protons of hypophosphite ions may occur as a result of exchange of the protons themselves and as a result of cross relaxation or relaxation of the second type. An estimate of the rate of exchange of protons between water and hypophosphite ions [188] showed that the lifetime of hypophosphite protons in this exchange is more than 100 hr, i.e., the rate of spin exchange measured in the present work is determined wholly by cross relaxation or relaxation of the second type.

It is interesting to note that in the experiment described above for measuring the rate of spin exchange an appreciable increase in the amplitude of the lines of hypophosphite protons was observed with saturation of the line of water protons in the first detector. In the first approximation, this increase was proportional to the decrease in magnetization of the water protons, i.e., $1-\cos\theta$, where θ is the angle of nutation of the magnetizations of the water protons.

The magnitude of the increase depends on the topography of the magnetic field in the first detector, the temperature, and the liquid flow velocity. Under optimal conditions, as a result of inversion of the magnetization of water protons there was an increase in the magnetization of the hypophosphite protons by a factor of 3. Analysis of the experimental results indicates that this effect is connected with pulsed nucleus-nucleus dynamic polarization in the first detector.

150

It is unusual that this occurs with a frequency of the oscillating field equal to the precession frequency of water protons. Polarization is not observed with an oscillating field with a frequency equal to the sum or difference of the resonance frequencies of water and hypophosphite protons.

APPENDIX

1. Change in Magnetization of Nuclei Under the Influence of a Resonance Rotating Magnetic Field

For simplicity, let us consider Bloch's equations, transformed by Wangness' method [185] into a system of coordinates with the x, y, and z axes rotating with a frequency ω together with the field H_1 about the z axis, along which the steady external field H_0 is directed. This system has the following form:

$$\left.\begin{aligned}
\frac{dM_x}{dt} + \frac{M_x}{T_{2n}} + (\omega_0 - \omega)\,M_y &= 0,\\
\frac{dM_y}{dt} + \frac{M_y}{T_{2n}} - (\omega_0 - \omega)\,M_x + \gamma H_1 M_z &= 0,\\
\frac{dM_z}{dt} + \frac{M_z}{T_{1n}} - \frac{M_0}{T_1} - \gamma H_1 M_y &= 0.
\end{aligned}\right\} \qquad (1)$$

The vector of the rotating field H_1 is directed along the x axis.

Here M_x, M_y, and M_z are the projections of the magnetization on the rotating axes x, y, and z, respectively, $M_0 = X_0 H_0$ is the equilibrium value of the magnetization in the field H_0, X_0 is the static nuclear magnetic susceptibility, $\omega_0 = \gamma H_0$ is the precession frequency of the nuclei in the field H_0, T_{1n} is the effective relaxation time, characterizing the decrease on nutation of the component M_z of the nuclear magnetization, T_{2n} is the effective relaxation time, characterizing the decrease on nutation of the components M_x and M_y of the nuclear magnetization, and T_1 is the natural relaxation time, characterizing the increase in the component M_z of the nuclear magnetization.

Let us find an expression for the projection of the magnetization on condition that there is accurate resonance tuning ($\omega = \omega_0$) in accordance with the general rule for the solution of systems of differential equations.

The first equation of the system

$$\frac{dM_x}{dt} + \frac{M_x}{T_{2n}} = 0 \qquad (2)$$

gives $M = C \exp(-t/T_{2n})$. Since at the first moment the magnetization of the nuclei is directed along the z axis, when $\omega = \omega_0$

$$M_z = 0.$$

For finding M_y and M_z it is necessary to solve the system

$$\left.\begin{aligned}
\frac{dM_y}{dt} + \frac{M_y}{T_{2n}} + \gamma H_1 M_z &= 0,\\
\frac{dM_z}{dt} + \frac{M_z}{T_{1n}} - \frac{M_0}{T_1} - \gamma H_1 M_y &= 0.
\end{aligned}\right\} \qquad (3)$$

This is an inhomogeneous system of linear differential equations of the 1st order. The solution equals the sum of the general solution of the corresponding homogeneous system and the particular solution of the inhomogeneous system.

Let us find the solution of the homogeneous system

$$\begin{aligned}
\frac{dM_y}{dt} &= -\frac{M_y}{T_{2n}} - \gamma H_1 M_z, \\
\frac{dM_z}{dt} &= +\gamma H_1 M_y - \frac{M_z}{T_{1n}}.
\end{aligned} \right\} \tag{4}$$

It will be in the form

$$\begin{aligned}
M_y &= A_1 e^{r_1 t} + A_2 e^{r_2 t}, \\
M_z &= B_1 e^{r_1 t} + B_2 e^{r_2 t}.
\end{aligned} \right\} \tag{5}$$

The characteristic equation for the system (4) has the form

$$\begin{vmatrix}
-\dfrac{1}{T_{2n}} - r & -\gamma H_1 \\[2mm]
+\gamma H_1 & -\dfrac{1}{T_{1n}} - r
\end{vmatrix} = 0 \tag{6}$$

or after expansion of the determinant

$$r^2 + \left(\frac{1}{T_{1n}} + \frac{1}{T_{2n}} \right) r + \gamma^2 H_1^2 + \frac{1}{T_{1n} T_{2n}} = 0. \tag{7}$$

By solving equation (7) we obtain

$$r_{1,2} = -\frac{1}{2T_{1n}} - \frac{1}{2T_{2n}} + \sqrt{\left(\frac{1}{2T_{2n}} - \frac{1}{2T_{1n}} \right) - \gamma^2 H_1^2}. \tag{8}$$

The values of the coefficients A and B may be found from the system

$$\begin{aligned}
\left(\frac{1}{T_{2n}} + r_{1,2} \right) A_{1,2} + \gamma H_1 B_{1,2} &= 0, \\
-\gamma H_1 A_{1,2} + \left(\frac{1}{T_{1n}} + r_{1,2} \right) B_{1,2} &= 0.
\end{aligned} \right\} \tag{9}$$

As the determinant of this system equals zero, it is possible to find only the ratio of the coefficients A and B from it. They equal the ratio of the cofactors of the corresponding terms of any line. Let us use the cofactors of the second line, then

$$\begin{aligned}
\frac{A_1}{B_1} &= \frac{\gamma H_1}{\dfrac{1}{T_{2n}} + r_1}, \\[3mm]
\frac{A_2}{B_2} &= -\frac{\gamma H_1}{\dfrac{1}{T_{2n}} + r_2}.
\end{aligned} \right\} \tag{10}$$

Hence

$$A_1 = C_1 \gamma H_1; \quad B_1 = -C_1 \left(\frac{1}{T_{2_n}} + r_1 \right),$$
$$A_2 = C_2 \gamma H_1; \quad B_2 = -C_2 \left(\frac{1}{T_{2_n}} + r_2 \right). \qquad (11)$$

By substituting these values in expression (5) we obtain the general solution of the system

$$M_y = \gamma H_1 (C_1 e^{r_1 t} + C_2 e^{r_2 t}),$$
$$M_z = -C_1 \left(\frac{1}{T_{2_n}} + r_1 \right) e^{r_1 t} - C_2 \left(\frac{1}{T_{2_n}} + r_2 \right) e^{r_2 t}. \qquad (12)$$

To find the particular solution of the inhomogeneous system we use the method of variation of arbitrary constants. Let $C_1 = C_1(t)$ and $C_2 = C_2(t)$. After substitution of the solutions (12) in the inhomogeneous system (3) we obtain a system of differential equations for the values of C

$$\frac{dC_1}{dt} e^{r_1 t} + \frac{dC_2}{dt} e^{r_2 t} = 0,$$
$$-\frac{dC_1}{dt} \left(\frac{1}{T_{2_n}} + r_1 \right) e^{r_1 t} - \frac{dC_2}{dt} \left(\frac{1}{T_{2_n}} + r_2 \right) = \frac{M_0}{T_1}. \qquad (13)$$

By solving this system algebraically, we obtain an expression for the derivatives of C

$$-\frac{dC_1}{dt} = \frac{M_0 e^{-r_1 t}}{T_1 (r_2 - r_1)},$$
$$\frac{dC_2}{dt} = \frac{M_0 e^{-r_2 t}}{T_1 (r_1 - r_2)}. \qquad (14)$$

By integration we obtain an expression for C (as a particular solution is being found, the arbitrary constants may be taken as equal to zero)

$$C_1(t) = \frac{M_0 e^{-r_1 t}}{T_1 r_1 (r_1 - r_2)},$$
$$C_2(t) = \frac{M_0 e^{-r_2 t}}{T_1 r_2 (r_2 - r_1)}. \qquad (15)$$

By substituting these values in expression (12), we obtain the required particular solution, which, after substitution of r_1 and r_2 from formula (8), has the form

$$M_y = -M_0 Z \gamma H_1 T_{2_n} T_{1_n} \frac{1}{T_1},$$
$$M_z = \frac{M_0 Z T_{1_n}}{T_1},$$
$$Z = \frac{1}{1 + \gamma^2 H_1^2 T_{1_n} T_{2_n}}. \qquad (16)$$

We write the general solution of the inhomogeneous equation

$$M_y = C_1 \gamma H_1 e^{r_1 t} + C_2 \gamma H_1 e^{r_2 t} - M_0 Z \gamma H_1 \frac{T_{1n}}{T_1} T_{2n'} \left.\begin{array}{c}\\\\\end{array}\right\}$$

$$M_z = -C_1 \left(\frac{1}{T_{2n}} + r_1\right) e^{r_1 t} - C_2 \left(\frac{1}{T_{2n}} + r_2\right) e^{r_2 t} + M_0 Z \frac{T_{1n}}{T_1}.$$

(17)

From the initial conditions when t = 0

$$M_{y(t=0)} = 0 = C_1 \gamma H_1 + C_2 \gamma H_1 - M_0 Z \gamma H_1 \frac{T_{1n} T_{2n}}{T_1}, \left.\begin{array}{c}\\\\\end{array}\right\}$$

$$M_{z(t=0)} = M_p = -C_1 \left(\frac{1}{T_{2n}} + r_1\right) - C_2 \left(\frac{1}{T_{2n}} + r_2\right) + M_0 Z \frac{T_{1n}}{T_1}.$$

(18)

We find C_1 and C_2:

$$C_1 = \frac{M_p + M_0 Z T_{2n} r_2 \frac{T_{2n}}{T_1}}{r_2 - r_1}, \left.\begin{array}{c}\\\\\\\\\end{array}\right\}$$

$$C_2 = \frac{M_p + M_0 Z T_{2n} r_1 \frac{T_{1n}}{T_1}}{r_1 - r_2}.$$

(19)

By substituting these values in system (17) we obtain the final expressions for M_y and M_z

$$M_y = -\frac{M_p \gamma H_1}{r_1 - r_2}(e^{r_1 t} - e^{r_2 t}) - M_0 Z \gamma H_1 T_{2n} \frac{T_{1n}}{T_1}\left(1 + \frac{r_2 e^{r_1 t} - r_1 e^{r_2 t}}{r_1 - r_2}\right),$$

(20)

$$M_z = \frac{M_p}{r_1 - r_2}\left[\left(\frac{1}{T_{2n}} + r_1\right)e^{r_1 t} - \left(\frac{1}{T_{2n}} + r_2\right)e^{r_2 t}\right] + \frac{M_0 Z}{r_1 - r_2}(r_2 e^{r_1 t} - r_1 e^{r_2 t})\frac{T_{1n}}{T_1} +$$

$$+ \frac{M_0 Z T_{2n} r_1 r_2}{r_1 - r_2}(e^{r_1 t} - e^{r_2 t})\frac{T_{1n}}{T_1} + M_0 Z \frac{T_{1n}}{T_1},$$

(21)

where

$$\frac{r_1 - r_2}{2} b = \sqrt{\left(\frac{1}{2T_{2n}} - \frac{1}{2T_{1n}}\right)^2 - \gamma^2 H_1^2}.$$

By substituting r_1 and r_2 from expression (8) in relation (21)

$$M_z = \left\{\left(M_p - M_0 Z \frac{T_{1n}}{T_1}\right)\left[\frac{e^{bt} - e^{-bt}}{2} + \frac{(e^{bt} - e^{-bt})(T_{1n} - T_{2n})}{4bT_{1n}T_{2n}}\right] + \right.$$

$$\left. + \frac{M_0(1 - Z)(e^{bt} - e^{-bt})}{2bT_1}\right\} e^{-\frac{t}{2}\left(\frac{1}{T_{1n}} + \frac{1}{T_{2n}}\right)} + M_0 Z \frac{T_{1n}}{T_1}.$$

(22)

2. Relation of Nuclear Resonance Signal Amplitude to Magnetization of Nuclei in Volume of Detector

The effect produced by the resonating nuclei on the detector circuit may be reduced to the appearance of a complex dynamic magnetic susceptibility

$$X = X' - iX''.$$

(1)

The oscillating magnetic field H directed along the axis of the coil of the circuit is equivalent to two fields of half the amplitude rotating in opposite directions

$$H = 2H_1 \cos \omega t = H_1(e^{i\omega t} + e^{-i\omega t}).$$

(2)

155

The interaction of the complex magnetic susceptibility with one of the rotating components of the oscillating field results in the appearance of a component of the vector of the magnetization of the nuclei at right angles to the external field

$$M_\perp = XHe^{\pm i\omega_0 t}. \tag{3}$$

(The sign plus or minus depends on the sign of the gyromagnetic ratio of the nuclei). By substituting in expression (3) X from relation (1) we obtain

$$M_\perp = 2X'H_1 e^{\pm i\omega_0 t} - 2X''H_1 e^{\pm i\left(\omega_0 t + \frac{\pi}{2}\right)}. \tag{4}$$

Let us denote the amplitude of the component $M_\perp$, rotating in phase with the component of the oscillating field by M_x and that rotating in quadrature with it by M_y, then from expression (4) it may be concluded that

$$X' = \frac{M_x}{2H_1}, \tag{5}$$

$$X'' = -\frac{M_y}{2H_1}. \tag{6}$$

The coil of the nuclear resonance detector is connected to a parallel nuclear resonance detector. The total resistance of this circuit is given by

$$z = \frac{z_L z_C}{z_L + z_C}, \tag{7}$$

where z_L and z_C are the inductive and capacative reactances of the circuit.

If we neglect the active resistance of the capacity, then

$$z_L = r + i\omega L_0, \tag{8}$$

$$z_C = -\frac{i}{\omega C}, \tag{9}$$

where r is the active resistance of the detector coil and L_0 is its inductance in the absence of resonating nuclei. The appearance in the working volume of the detector of a dynamic magnetic susceptibility X changes the inductance of the circuit

$$L = L_0 (1 + 4\pi\eta X), \tag{10}$$

where η is the filling factor of the coil. Thereupon the resistance of the inductive branch of the circuit changes by the value

$$\Delta z_L = i\omega L_0 4\pi\eta X. \tag{11}$$

Let us find the change in the total resistance of the circuit

$$\Delta z = \frac{\partial z}{\partial z_L} \Delta z_L. \tag{12}$$

From expression (7)

$$\frac{\partial z}{\partial z_L} = \frac{z_C^2}{(z_L + z_C)^2} \cdot \tag{13}$$

By substituting in formula (13) expressions (8) and (9) we obtain

$$\frac{\partial z}{\partial z_L} = \frac{\dfrac{1}{\omega^2 C^2}}{\left(r + i\omega L_0 - \dfrac{i}{\omega C}\right)^2} \cdot \tag{14}$$

When the circuit is tuned for resonance with the precession frequency of the nuclei

$$\frac{\partial z}{\partial z_L} = -\frac{1}{\omega_0^2 C^2 r^2}, \tag{15}$$

where

$$\omega_0 = \frac{1}{\sqrt{L_0 C}} \cdot$$

By substituting the values from (15) and (11) in expression (12) we obtain

$$\Delta z = -4\pi \eta Q^2 \omega_0 L_0 \, (X'' + iX'), \tag{16}$$

where $Q = \omega_0 L_0 / r$ is the quality factor of the circuit.

If the oscillations in the circuit are excited by an oscillator with a constant current I, the change in the potential in the circuit at resonance, i.e., the signal amplitude

$$A = I\,\Delta z. \tag{17}$$

By substituting Δz from expression (16) we obtain

$$A = -4\pi \eta Q^2 \omega_0 L_0 I \, (X'' + iX'). \tag{18}$$

From the definition of inductance

$$L_0 I Q = 2H_1 NS, \tag{19}$$

where N is the number of turns of the coil and S the area of a turn.

By substituting L_0 from expression (19) in expression (18) we obtain

$$A = -8\hbar \eta NSQ\omega_0 H_1 \, (X'' + iX'). \tag{20}$$

From expression (20) it follows that the amplitude of the signal in phase with the current (absorption)

$$A_a = -8\pi \eta NSQ\omega_0 H_1 X'', \tag{21}$$

while the amplitude of the signal in quadrature with the current(dispersion)

$$A_d = -8\pi\eta NSQ\omega_0 H_1 X'. \tag{22}$$

By substituting the value of X" from expression (6) in formula (21) we obtain the relation of the absorption signal amplitude to the magnetization of the nuclei in the detector

$$A_a = 4\pi\eta QNS\omega_0 M_y. \tag{23}$$

3. Relation of Absorption Signal Amplitude to the Nonuniformity of the Field in the Volume of the Flow Detector

As was shown in Section 3.2, the nonuniformity of the field in the detector at right angles to the flow of the liquid produces dephasing of the precessing magnetic moments, making $T_{2n} \approx 2/\gamma\Delta H_\perp \ll T_{1n}$. Let us adopt the symbols $v_a/2qT_{2n} = (v_a/4q)\gamma\Delta H_\perp = a$; $(v_a/q)\gamma H_1 = \theta$.

When $Z \ll 1$ and $(v_a/qT_1) \ll 1$, expression (5.3) for the signal amplitude has the form:

if $\theta \geq a$, then

$$A = \frac{A_M}{\theta}\left[1 - e^{-a}\left(\cos\sqrt{\theta^2 - a^2} + \frac{\sin\sqrt{\theta^2 - a^2}}{\sqrt{\frac{\theta^2}{a^2} - 1}}\right)\right]; \tag{1}$$

when $\theta \leq a$ the transverse nonuniformity of the field in the detector ($\Delta H_\perp$) is greater than the width of the nuclear resonance line ($4H_1$) and only part of the cross section of the working volume participates in the resonance effect at one time and therefore the factor $4H_1/\Delta H_\perp = \theta/a$

$$A = \frac{A_M}{a}\left[1 - e^{-a}\left(\frac{e^{a\sqrt{1-\frac{\theta^2}{a^2}}} + e^{-a\sqrt{1-\frac{\theta^2}{a^2}}}}{2} + \frac{e^{a\sqrt{1-\frac{\theta^2}{a^2}}} - e^{-a\sqrt{1-\frac{\theta^2}{a^2}}}}{2\sqrt{1-\frac{\theta^2}{a^2}}}\right)\right]. \tag{2}$$

a) $a \gg 1$. From expression (1) it follows that when $\theta \geq a$ the maximum amplitude $A_{max} = (A_M/a)[1 - e^{-a}(1 + a)]$ corresponds to $\theta = a$.

In expression (2), when $a \gg 1$ and $\theta < a$ it is possible to neglect the exponents with negative indices and then

$$A = \frac{A_M}{a}\left[1 - \frac{e^{a\left(\sqrt{1-\frac{\theta^2}{a^2}}-1\right)}}{2}\left(1 + \frac{1}{\sqrt{1-\frac{\theta^2}{a^2}}}\right)\right]. \tag{3}$$

If $\theta < 0.7a$, then by using the expansions

$$\sqrt{1 - \frac{\theta^2}{a^2}} = 1 - \frac{\theta^2}{2a^2} \quad \text{and} \quad \frac{1}{\sqrt{1-\frac{\theta^2}{a^2}}} = 1 + \frac{\theta^2}{2a^2},$$

instead of expression (3) we obtain

$$A = \frac{A_M}{a} \left[1 - e^{-\frac{\theta^2}{2a^2}} \left(1 + \frac{\theta^2}{4a^2} \right) \right] . \tag{4}$$

From this expression it follows that the maximum amplitude

$$A_{max} = \frac{A_M}{a} \tag{5}$$

when

$$\frac{\theta^2}{2a^2} > 5. \tag{6}$$

Thus, when $(v_a/q)\gamma\Delta H_\perp \gg 1$ the maximum signal amplitude

$$A_{max} = \frac{A_M}{\frac{v_a \gamma \Delta H_\perp}{4q}} . \tag{7}$$

b) $a \ll 1.$ When $\theta \gg a$ from expression (1) we obtain $A = (A_M/\theta) (1 - \cos\theta)$. Hence $A_{max} = 0.7 \, A_M$.

When $\theta < a$ from condition (2), by using the expansions $e^x = 1 + x$ and $1/\sqrt{1-x^2} = 1 + (x^2/2)|x < 0.7|$, we obtain

$$A = \frac{A_M}{a} \left\{ 1 - (1-a) \left(\left[1 + a \left(1 - \frac{\theta^2}{2a^2} \right) \right] \left(1 + \frac{\theta^2}{4a^2} \right) - \left[1 - a \left(1 - \frac{\theta^2}{2a^2} \right) \right] \frac{\theta^2}{4a^2} \right) \right\} \tag{8}$$

or

$$A = A_M \left[a + \frac{\theta^4}{4a^4} (1-a) \right]$$

From this expression it is evident that when $\theta < 0.7$, A increases with an increase in θ.

When $\theta = a$, from expression (1) we have

$$A = \frac{A_M}{a} [1 - e^{-a} (1+a)]. \tag{9}$$

TABLE 1.II

$a = \dfrac{v_a \gamma \Delta H_\perp}{4q}$	0.1	0.5	1	2	3	10
$\dfrac{A_{max}}{A_M}$	0.7	0.55	0.45	0.33	0.26	0.1
$\theta_{opt} = \dfrac{v_a}{q} \gamma H_{opt}$	2.33	2.5	2.5	3	3	8

By substituting $e^{-a} = 1-a$, we obtain $A = A_M a$, i.e., in the case when $\theta < a$, $A \ll A_M$.

Thus, if $v_a \gamma \Delta H_\perp / 4q \ll 1$, the maximum signal amplitude $A_{max} = 0.7 A_M$ when $\theta_{opt} = 2.33$

c) a = 1. In this case expression (1) and (2) were investigated numerically. The relation of A_{max} and θ_{opt} to a obtained is given in Table 1.II.

The result shows that when $v_a \gamma \Delta H_\perp / 4q > 1$, with an error of less than 10% the signal amplitude may be represented as follows:

$$A_{max} = \frac{A_M}{\dfrac{v_a \gamma \Delta H_\perp}{4q} + 1} \, . \tag{10}$$

4. Change in Magnetization of Nuclei Under the Influence of a Rotating Magnetic Field with a Frequency Different from the Precession Frequency of the Nuclei

In order to find an expression for the projection of the magnetization under conditions different from resonance conditions it is necessary to solve Bloch's system of equations when $\omega \neq \omega_0$. After substitution of $\omega_0 - \omega = \Delta\omega$, with the condition $H_0 \ll H_p$, this system has the form:

$$\left.\begin{aligned}
\frac{dM_x}{dt} &= -\frac{M_x}{T_{2n}} - \Delta\omega M_y, \\
\frac{dM_y}{dt} &= -\frac{M_y}{T_{2n}} + \Delta\omega M_x - \gamma H_1 M_z, \\
\frac{dM_z}{dt} &= \gamma H_1 M_y - \frac{M_z}{T_{1n}} \, .
\end{aligned}\right\} \tag{1}$$

We find the result according to the general rule for the solution of linear systems of equations. The determinant of the characteristic equation will be

$$\begin{vmatrix}
-\dfrac{1}{T_{2n}} - r & -\Delta\omega & 0 \\[2mm]
\Delta\omega & -\dfrac{1}{T_{2n}} - r & -\gamma H_1 \\[2mm]
0 & \gamma H_1 & -\dfrac{1}{T_{1n}} - r
\end{vmatrix} \tag{2}$$

The characteristic equation has the form

$$\left(-\frac{1}{T_{2n}} + r\right)^2 \left(\frac{1}{T_{1n}} + r\right) - \gamma^2 H_1^2 \left(\frac{1}{T_{2n}} + r\right) - \Delta\omega^2 \left(\frac{1}{T_{1n}} + r\right) = 0. \tag{3}$$

For simplicity we assume that $T_{1n} = T_{2n} = T$. This is valid with a low transverse nonuniformity of the field. The roots of the characteristic equation will be as follows:

$$\left.\begin{aligned}
r_1 &= -\frac{1}{T} \, , \\
r_{2,3} &= -\frac{1}{T} \pm i \sqrt{\gamma^2 H_1^2 + \Delta\omega^2} .
\end{aligned}\right\} \tag{4}$$

160

The next sequence of operations is analogous to that given in Appendix 2. After solution of the determinant, in order to find the coefficients and determine the arbitrary constants from the initial conditions, we obtain an expression for the projection of the magnetization when the frequency is not tuned exactly to resonance:

$$M_x = M_p e^{-\frac{t}{T}} \frac{\Delta\omega}{\gamma H_1 \left(\frac{\Delta\omega^2}{\gamma^2 H_1^2} + 1\right)} \left[1 - \cos t \sqrt{\gamma^2 H_1^2 + \Delta\omega^2}\right], \tag{5}$$

$$M_y = \frac{M_p e^{-\frac{t}{T}}}{\sqrt{\frac{\Delta\omega^2}{\gamma^2 H_1^2} + 1}} \sin t \sqrt{\gamma^2 H_1^2 + \Delta\omega^2}, \tag{6}$$

$$M_z = M_p e^{-\frac{t}{T}} \left[1 - \frac{1 - \cos t \sqrt{\gamma^2 H_1^2 + \Delta\omega^2}}{\frac{\Delta\omega^2}{\gamma^2 H_1^2} + 1}\right]. \tag{7}$$

LITERATURE

1. Suryan, G., Proc. Indian Acad. Sci., A33: 107 (1951).
2. Dennis, P. M., Béné, G. J., and Exterman, R. C., Arch. Sci., 5: 32 (1952).
3. Bloom, A. L., and Shoolery, J. N., Phys. Rev., 90: 358 (1953).
4. Sherman, C., Phys. Rev., 93: 1429 (1954).
5. Sherman, C., Rev. Sci. Instr., 30: 568 (1959).
6. Gaussen, A. Z., Z. Naturforsch., a10: 54 (1955).
7. Mitchell, A. M., and Phyllips, G., Brit. J. Appl. Phys., 7: 67 (1956).
8. Hrynkiewicz, A. Z., and Waluga, T., Acta Phys. Polon., 16, (5)381: (1957).
9. Antonowicz, K., Bull. Acad. Polon. Sci. Cl.III,Vol.5, No.11(1957).
10. Antonowica, K., Bull. Acad. Polon. Sci. Cl. III, Vol. 5, No. 8(1957).
11. Zhernovoi, A. I., Egorov, Yu. S., and Latyshev, G. D., Pribory i Techn. Eksperim., No. 5,73(1958).
12. Zhernovoi, A. I., Egorov, Yu. S., and Latyshev, G. D., Inzh.-Fiz. Zh., Vol. 1, 95 (1958).
13. Skripov, F. I., Dokl. Akad. Nauk SSSR, 121: 998(1958).
14. Zhernovoi, A. I., and Latyshev, G. D., Vestn. Akad. Nauk KazSSR, No. 5, (1960).
15. Benoit, H., and Ottavi, H., Compt. Rend., 249: 83(1959).
16. Benoit, H., and Ottavi, H., Compt. Rend. , 250: 2708(1960)
17. Ekaterinin, V. V., Zhernovoi, A. I., and Yakovlev, G. I., Izv. Akad. Nauk. KazSSR, Vol. 6 (1963).
18. Zhernovoi, A. I. Egorov, Yu. S., and Latyshev, G. D., Pribory i Tekhn. Eksperim., No. 5, 71 (1958).
19. Zhernovoi, A. I., Egorov, Yu. S., and Latyshev, G. D., Inz.-Fiz. Zh. 1(9): 123 (1958).
20. Zhernovoi, A. I., Egorov, Yu. S., and Latyshev, G. D., Izv. Akad. Nauk SSSR, Ser. Fiz.,Vol. 22, No.8(1859).
21. Wilking, S., Z. Physik, 157: 384 (1959).
22. Benoit, H. Grivet, P., and Cuibe, L., Compt. Rend., 246: 3609(1958).
23. Benoit, H., et al., J. Phys. Radium, 19: 905(1958).
24. Benoit, H., Grivet, P., and Ottavi, H., Compt. Rend., 247: 1985(1958).
25. Benoit, H., Grivet, P., and Ottavi, H., Compt. Rend., 248: 220 (1959).
26. Zhernovoi, A. I., Geofiz. Priborostr., No. 6, 66 (1960).
27. Hennequin, J., Compt. Rend., 250: 2711 (1960).
28. Benoit, H., and Ottavi, H., Compt. Rend. 249: 73 (1959).
29. Benoit, H., and Hennequin, J., Compt. Rend., 248: 1991(1959).
30. Fric, C., Compt. Rend., 249: 80(1959).
31. Benoit, H., and Fric, C., Compt. Rend., 249: 537 (1959).
32. Fric, C., Compt. Rend. 250: 2353(1960).
33. Benoit, H., J. Phys. Radium, 21: 212(1960).
34. Zhernovoi, A. I., Geofiz. Priborostr., No. 6, 59 (1960).
35. Zhernovoi, A. I., Pribory i Tekhn. Eksperim., No. 5, 12 (1961).
36. Zhernovoi, A. I., Authors'Certificate No. 131517, with priority from February 4, 1959.
37. Zhernovoi, A. I., Authors' Certificate No. 125906, with priority from May 18, 1959.
38. Zhernovoi, A. I., Authors' Certificate No. 143567, with priority from February 24, 1959.
39. Zhernovoi, A. I., Priborostroenie, No. 6 (1960).
40. Singer, J. R. Electronics 33: (14): 71 (1960).
41. Singer, J. R., J. Appl. Phys., 31: 125 (1960).
42. Singer, J. R., Science, 130: 1652 (1959).
43. Hrynkiewicz, A. Z., Acta Phys. Polon., 17(5): 353 (1958).
44. Zhernovoi, A. I., Authors' Certificate No. 128669, with priority from September 28, 1959.

45. Zhernovoi, A. I., Inz.-Fiz. Zh., 4:91(1961).

46. Zhernovoi, A. I., and Latyshev, G. D., Izv. Akad. Nauk SSSR, Ser. Fiz., 22:993 (1958).

47. Zhernovoi, A. I., Inz.-Fiz. Zh., 5:64 (1962).

48. Zhernovoi, A. I., Authors' Certificate No. 133262, February 13, 1959.

49. Ottavi, H., Arch. Sci., 14:360 (1961).

50. Ramsey, N. F., and Pound, R. V., Phys. Rev., 87:278 (1951).

51. Overhauser, A. W., Phys. Rev., 91:476 (1953).

52. Overhauser, A. W., Phys. Rev., 92:411 (1953).

53. Abraham, A., Phys. Rev., 98:1729 (1955).

54. Abraham, A., and Proctor, W. C., Compt. Rend., 246: 2253 (1953).

55. Garver, T. R. and Slichter, C. P., Phys. Rev., 102:975 (1956).

56. Bennett, L. H., and Torry, H. C., Phys. Rev., 108:499 (1957).

57. Abraham, A., Cambrisson, I., and Solomon, I., Compt. Rend., 245:157 (1957).

58. Bennet, M., and Servor-Garvin, M., Arch. Sci., 13: 629 (1961).

59. Zhernovoi, A. I., Latyshev, G. D., and Sergeev, A. G., Pribori i Tekhn. Eksperim., No. 2, 69 (1957).

60. Zhernovoi, A. I., and Latyshev, G. D., Vestn. Akad. Nauk KazSSR, No. 5, 74 (1959).

61. Skripov, F. I., Proceedings of a Conference on Paramagnetic Resonance (Kazan', 1960).

62. Zhernovoi, A. I., Rukhin, A. B., and Stakhov, V. V., Izv. Akad. Nauk SSSR, Ser. Fiz., 27(7):947 (1963).

63. Purcell, E. M., and Pound, R. V., Phys. Rev., 81:278 (1951).

64. Packard, M., and Varian, R., Phys. Rev., 93:941 (1954).

65. Hrynkiewicz, A. Z., Waluga, T., and Zapalski, G., Acta Phys. Polon., 17(1):71 (1958).

66. Hrynkiewicz, A. Z., Waluga, T., and Zapalski, G., Arch. Sci., 2:190 (1958).

67. Zhernovoi, A. I., and Latyshev G. D., Vestn. Akad. Nauk KazSSR, No. 10 (1963).

68. Hahn, E. L., Phys., Rev., 77:297 (1950).

69. Hahn, E. L., Phys. Rev., 80:580 (1950).

70. Das, T. P., and Saha, A. K., Phys. Rev., 93:749 (1954).

71. Hahn, E. L., Phys. Today, 6(11):4 (1953).

72. Torrey, H. C. Phys. Rev., 76:1059 (1949).

73. Torrey, H. C. Phys. Rev., 85:365 (1952).

74. Bloch, F., and Siegert, A., Phys. Rev., 57:522 (1940).

75. Wilking, S. Z. Physik, 157:401 (1959).

76. Bloch, F., Phys. Rev., 70:460 (1946).

77. Bloch, F., Hansen, W. W., and Packard, M. K. Phys. Rev., 70:474 (1946).

78. Powles, G., Proc. Phys. Soc., 71:497 (1958).

79. Gvozdover, S. D., and Magazanik, A. A., Zh. Eksperim. i Teor. Fiz., 20:705 (1950).

80. Drain, L. E., Proc. Phys. Soc., A62:301 (1949).

81. Benoit, H., Ann. Phys., 4:1439 (1959).

82. Fric, C., Ann. Phys., 5:1501 (1960).

83. Hennequin, J., Ann. Phys. 6:946 (1961).

84. Hirshel, L., and Libelo, L., J. Appl. Phys., 32(3):1401 (1961).

85. Hirshel, L., and Libelo, L., J. Appl. Phys., 33(5):1895 (1962).

86. Herms, W., Ann. Physik, 8:280 (1961).

87. Egorov, Yu. S., and Latyshev, G. D. Pribory i Tekhn. Eksperim. No. 2, 80 (1956).

88. Packard, M. E., Rev. Sci. Instr., 19: 435 (1950).

89. Levinthal, E. C., Phys. Rev., 78:204 (1950).

90. Proctor, W. G., Phys. Rev., 79:35 (1950).

91. Weaver, H. E. Phys. Rev., 89:923 (1953).

92. Blombergen, N., and Pound, R. V. Phys. Rev., 95(1):8 (1954).

93. Vladimirskii, K. V., Zh. Eksperim. i Teor. Fiz., 33:352 (1957).

94. Gordon, Z. P., Zeiger, J., and Townes, C. H., Phys. Rev. 99:1265 (1955).

95. Bloch, F., and Siegert, A., Phys. Rev., 93:1241 (1954).

96. Anderson, Phys. Rev., 102:151 (1956).

97. Seiden, Compt. Rend., 240: 2228 (1955).

98. Hopkins, N. I., Rev. Sci. Instr., 20: 401 (1949).

99. Pound, R. V., and Knight, W. D., Rev. Sci. Instr., 21: 219 (1950).

100. Pound, R. V., and Knight, W. D., Rev. Sci. Instr., 21: 942 (1950).

101. Knoebel, H. W., and Hahn, E. L., Rev. Sci. Instr., 22: 904 (1951).

102. Laroche, A., Nature, No. 3244, 320 (1955)

103. Yagola, G. K., et al., Izmerit, Tekhn., 6: 9 (1955).

104. Leont'ev, N. I., Zh. Eksperim. i Teor. Fiz., 28(1): 77 (1955).

105. Denisov, Yu. N., Pribory i Tekhn. Eksperim., No. 5, 67 (1958).

106. Gertsiger, L. N., Pribory i Tekhn. Eksperim., No. 2, 33 (1959).

107. Denisov, Yu. N., Pribory i Tekhn. Eksperim., No. 1, 96 (1959).

108. Packard, M. E. Rev., Sci. Instr., 19: 435 (1948).

109. Thomas, H.A., Driscoll, R. L., and Hipple, J. A., J. Res. Nat. Bur. Standards, 44: 569 (1950).

110. Thomas. H. A. Electronics, 24: 114 (1952).

111. Lindstrom, G., ArkivFys., 4: 9151 (1952).

112. Shpigel', I. S., Raizer, M. D., and Maé, É. A., Radiotekhn. i Elektrotekhn., 1: 1515 (1956).

113. Egorov, Yu. S., Latyshev, G. D., and Trulev, Yu. N., Pribory i Tekhn. Eksperim., No. 5, 41 (1957).

114. Winter, I., et al., Arch. Sci., 7: 23 (1954).

115. Winter, I., et al., Compt. Rend., 239: 803 (1954).

116. Brown, R. M., and Purcell, E. M., Phys. Rev., 75: 1262 (1949).

117. Brown, R. M., Phys. Rev., 78: 530 (1950).

118. Manus, C., et al., Helv. Phys. Acta, 28(7): 617 (1955).

119. Cagnac, B., Helv. Phys. Acta, 28(7): 626 (1955).

120. Packard, M., and Varyan, R., Bull. Amer. Phys. Soc. Vol.28, No. 7 (1953).

121. Collerman, F. Phys. Rev., 93: 941 (1954).

122. Mel'nikov, A. V., et al., Zh. Tekhn. Fiz., 28: 900 (1958).

123. Amosov, A. P., Rotshtein, A. Ya., and Tsirel', V. S., Geofiz. Priborostr., No. 6, 33 (1960).

124. Packard, M. E., Rev. Sci. Inst., 19: 435 (1948).

125. Thomas, H. A., Driscoll, R. L., and Hipple, J. A. J. Res. Nat. Bur. Standards, 44: 569 (1950).

126. Thomas, H. A., Electronics, 25: 114 (1952).

127. Morozov, A. A., Mel'nikov, A. V., and Skripov, F. I., Izv. Akad. Nauk SSSR, Ser. Fiz., 23: 1141 (1958).

128. Egorov, Yu. S., et al., Izv. Akad. Nauk. SSSR, Ser. Fiz., 23(2): 244 (1958).

129. Inozemtsev, K. V., and Latyshev, G. D., Izv. Akad. Nauk SSSR, Ser. Fiz., 13: 453 (1949).

130. Symonds, J. L., Rept. Progr. Phys., 18: 83 (1955).

131. Cole, R. H. Rev. Sci. Instr., 9: 215 (1938).

132. Kanakhovich, Yu. Ya., Latyshev, G. D., and Tsimbalin, V. V., Izv. Akad. Nauk SSSR, Ser. Fiz., 13: 456. (1949).

133. Thomas, H. A., Driscoll, R. L. and Hipple, J. A. J. Res. Nat. Bur. Standards, 44: 569 (1950).

134. Chang, W. J., and Rosenblum, S., Rev., Sci. Instr., 15: 75 (1946).

135. Lauritsen, C. C. Rev. Sci. Instr., 19: 916 (1948).

136. Birebent, R., Compt. Rend., 240: 1064 (1955).

137. Birebent, R., Compt. Rend., 241: 368 (1955).

138. Cress, E. C. Rev. Sci. Instr., 18: 77 (1947).

139. Adams, C. D. Dussel, R. W., and Towslery, P. E., Rev. Sci. Instr., 21: 69 (1950).

140. Peregud, P. B., Pribory i Tekhn. Eksperim., No. 3, 64 (1957).

141. Meek, J. H., and Hector, F. S., Canad. J. Phys., 33: 364 (1955).

142. Rose, D. C., and Bloom, J. N., Canad. J. Res., A28: 153 (1950).

143. Peregud, P. B., Pribory i Tekhn. Eksperim., No. 5, (1958).

144. Kovrigin, O. D., and Latyshev, G. D., Inz.-Fiz. Zh., No. 11, 92 (1958).

145. Pearson, G. L., Rev. Sci. Instr., 19: 263 (1948).

146. Versois, P. L., Nature, No. 3247, 434 (1955).

147. Mason, W. R., Newitt, W. H., and Wick, R. F., J. Appl. Phys., 24: 166 (1953).

148. Sominskii, M. S., Vestn. Akad. Nauk SSSR, 27:48 (1957).

149. Vasilevskaya, D. P., and Denisov, Yu. N., Pribory i Tekhn. Eksperim., No. 3, 144 (1959).

150. Voeikov, D. D., Pribory i Tekhn. Eksperim., No. 4, 100 (1959).

151. Sus, A. N., and Bogdanov, N. N., Pribory i Tekhn. Eksperim., No. 5, 117 (1959).

152. Regel', A. R., Semiconductor Instruments for Measuring Magnetic Field Strength (Moscow-Leningrad: Izd. Akad. Nauk SSSR, 1956).

153. Chirkov, A. K., Pribory i Tekhn. Eksperim., No. 2, 36 (1957).

154. Garstens, M. A., and Ryan, A. H., Phys. Rev., 81:888 (1951).

155. Fait, Z., Czech. J. Phys., 9:218 (1957).

156. Gabillard, R., Arch. Sci., 9:84 (1956).

157. Denisov, Yu. N., Pribory i Tekhn. Eksperim., No. 1, 82 (1960).

158. Leont'ev, N. I., and Udovichenko, Yu. K., Authors' Certificate No. 193117, with priority from April 15, 1958.

159. Leont'ev, N. I., Pribory i Tekhn. Eksperim., No. 2, 93 (1960).

160. Leont'ev N. I., and Udovichenko, Yu. K., Authors' Certificate No. 19316, with priority from April 15, 1958.

161. Leont'ev, N. I., Pribory i Tekhn. Eksperim., No. 1, 78 (1960).

162. Dubovoi, L. V. Shvets, O. M., and Ovchinnikov, S. S., Pribory i Tekhn. Eksperim., No. 3, 106 (1960).

163. Benoit, H., and Ottavi, H., Compt. Rend., 250:2886 (1960).

164. Béné, G., Denis, P. M., and Exterman, R. C., Helv. Phys. Acta, 24:663 (1951).

165. Spektor, S. A., Authors' Certificate No. 110291, December 31, 1955.

166. Spektor, S. A., Nauchno-Tekhn. Inform. Byul. Leningrad. Politekhn. Inst., No. 11, 54 (1957).

167. Kremlevskii, P. P., Flowmeters, p. 400, (Moscow-Leningrad:Mashgiz, 1955).

168. Nikitin, B. I., Priborostroenie, No. 7, 13 (1956).

169. Andrew, E. R., Nuclear Magnetic Resonance, p. 124, (Moscow:Izd. IL, 1957).

170. Torrey, H. C., Phys. Rev., 75:1326 (1949).

171. Hennel, I. W., and Hrynkiewicz, A. Z., Arch. Sci., 11: 238(1958).

172. Hennel, I. W., et al., Arch. Sci., 11:243 (1958).

173. Hahn, E. L., Phys. Rev., 76:145 (1949).

174. Carr, N. Y., and Purcell, E. M., Phys. Rev., 94:630 (1954).

175. Torne, E. C., Compt. Rend., 250:512 (1960)

176. Chose, T., Nuovo Cimento, 5, 1771 (1957).

177. Pomerantsev, N. M., Usp. Fiz. Nauk, No. 1, 65 (1958).

178. Korepanov, V. D., Doutov, R. A., and Fadeev, V. M., Zh. Eksperim. i. Teor. Fiz., No. 1, 308 (1959).

179. Solomon, I., J. Phys. Radium, 20:788 (1959).

180. Conger, R. L., and Selwood, F. W., J. Chem. Phys., 20:383 (1952).

181. Chiarotti, G., and Guilotto, L., Phys. Rev., 93: 1241 (1954).

182. Manus, C., et al., Compt. Rend., 238:1315 (1954).

183. Samoilov, O. Ya., Structure of Aqueous Solutions of Electrolytes and Hydration of Ions (Moscow: Izd. Akad. Nauk SSSR, 1957).

184. Rogers, E. H., and Staub, A. H., Phys. Rev., 76:980 (1949).

185. Rogers, E. H., and Staub, A. H., Helv. Phys. Acta, 23:63 (1950).

186. Levintal, E. C., Phys. Rev., 78:204 (1950).

187. Alder, F., and Yu, F. C., Phys. Rev., 82:105 (1951).

188. Brodskii, A. I., and Sulema, L. V., Dokl. Akad. Nauk SSSR, 88:1277 (1952).

189. Bloch, F., and Siegert, A., Phys. Rev., 57:522 (1940).

190. Bloch, F.,Hansen, W. W., and Packard, M. E., Phys. Rev., 69:127 (1946).

191. Cuilotto, L. Arch. Sci., 9:212 (1956).

192. Wangness, R. K., Am. J. Phys., 21:275 (1953).

193. Agroskin, I. I., Dmitriev, G. D., and Pikalov, F. I., Hydraulics (Moscow:Gosenergoizdat, 1954).

194. Zhernovoi, A. I., Inzh.-Fiz. Zh., 5, No. 5 (1962).